Reimagining Psychoanalytic Theory in the Climate Polycrisis

Reimagining Psychoanalytic Theory in the Climate Polycrisis re-visions psychoanalysis by considering indigenous philosophy and the work of Giorgio Agamben.

Ryan LaMothe argues that the origins of psychoanalysis have tacitly produced, with some qualifications, not only dismissive attitudes and relations toward so-called "primitive" peoples and their philosophies, but also depersonalization of other species. Revisiting key psychoanalytic concepts and drawing on Winnicott, Bollas, and Lear, LaMothe sheds new light on notions of subjectivity, psychosocial development, trauma, transference, resistance, therapy, and analytic love and hope. Relying on the philosophy of Giorgio Agamben and Indigenous philosophies, this book aims to cross the divide by reimagining psychoanalytic developmental theory and concepts.

Reimagining Psychoanalytic Theory in the Climate Polycrisis will be essential reading for psychoanalysts and psychotherapists in practice and in training.

Ryan LaMothe is a professor of pastoral care and counselling. He has written extensively in the areas of psychoanalysis, political philosophy and theology, and psychology of religion. Recent work includes *A Political Psychoanalysis for the Anthropocene Age* (Routledge, 2023).

"LaMothe's creativity critiquing and reimagining theories and practices offers hope for a world at risk. His argument strengthens a much needed and unfolding paradigm of care that bridges the human/nonhuman divide. The book is destined to become foundational in psychoanalytic theory, training, and practice. Therapists, cultural critics, and individuals who love and care are invited to explore ethical ways of living on a species-rich earth."

Jaco J. Hamman, Professor of Religion, Psychology, and Culture, Vanderbilt University

"In the context of the climate polycrisis and its attendant anxieties, Ryan LaMothe offers a powerful critique of the way humans have seen themselves as separate from, and superior to, other species, to the detriment of human development and the survival of the planet. In a brilliant and sophisticated exploration that draws on psychoanalysis, philosophy, ethics, indigenous wisdom, and clinical experience, this book offers both a solemn warning and a path toward healing grounded in an ethic of care that embraces the suffering of all species. Scholars, clinicians, and students will all benefit from their encounter with this moving work that offers a vision for a future where all inhabitants of the earth might survive and thrive together."

Lisa Cataldo, psychoanalyst and Associate Professor of Counseling, Fordham University

"In this intensely ambitious book Ryan LaMothe performs an overdue and urgent Augean labor in clearing the messy ontological decks in order to render psychoanalysis more useful for coming to grips sanely with the climate crisis and all its implications. Nothing is sacred and everything is. This battered biosphere can stand all the 'anarchic care' that we can muster. Highly and ungovernably recommended."

Kurt Jacobsen and **David Morgan**, co-editors of *Free Associations*

"In the midst of our intensifying ecological breakdown, LaMothe critiques the Western ontological rift between the human and everything else—a destructive boundary that psychoanalysis has absorbed. Here, guided by Agamben and indigenous philosophies, LaMothe offers hope for a viable future by exploring ontoepistemologies that refuse to be incorporated into this radical separation. He imagines a psychoanalysis that could contribute to an ecologically sane future. This is a book for our time!"

Kristin Fiorella, Psy.D., psychoanalyst on the faculties of the Psychoanalytic Association of Northern California and the San Francisco Center for Psychoanalysis, and editor of *Child and Adolescent Psychoanalysis in Times of Crisis: War, Pandemic, and Climate Change*

Reimagining Psychoanalytic Theory in the Climate Polycrisis

Thinking with Giorgio Agamben and Indigenous Philosophies

Ryan LaMothe

LONDON AND NEW YORK

Designed cover image: © Getty Images

First published 2026
by Routledge
4 Park Square, Milton Park, Abingdon, Oxon OX14 4RN

and by Routledge
605 Third Avenue, New York, NY 10158

Routledge is an imprint of the Taylor & Francis Group, an informa business

British Library Cataloguing-in-Publication Data
A catalogue record for this book is available from the British Library

ISBN: 978-1-032-85411-3 (hbk)
ISBN: 978-1-032-85410-6 (pbk)
ISBN: 978-1-003-51801-3 (ebk)

DOI: 10.4324/9781003518013

Typeset in Times New Roman
by Taylor & Francis Books

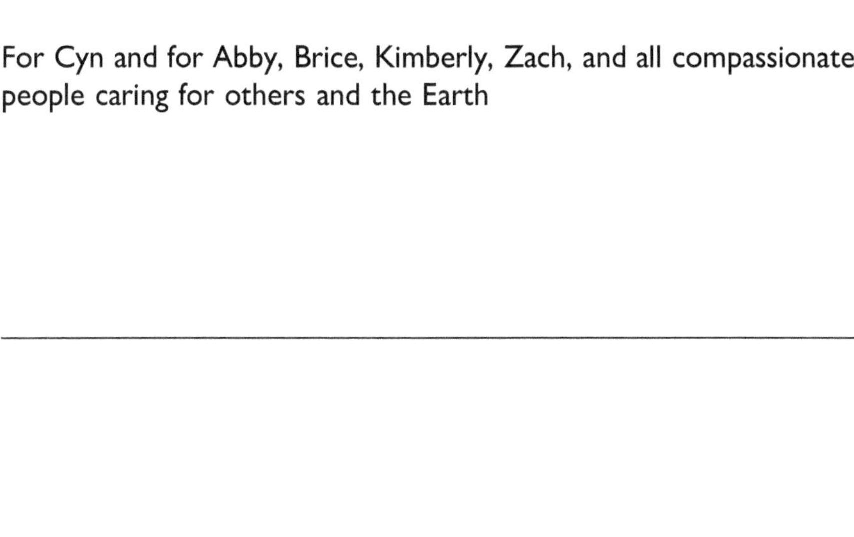

For Cyn and for Abby, Brice, Kimberly, Zach, and all compassionate people caring for others and the Earth

Contents

Acknowledgments

Every book represents a journey. Some days the road is smooth and straight and other times filled with obstacles. However arduous, the journey in writing a book comes with the joys and pleasures of meeting new people, observing and sharing surprising vistas with traveling companions, and obtaining respites with hospitable guides. In preparation for this trek, I have been helped and supported by conversations with colleagues, as well as putting forth ideas at various conferences and groups. Robert Dykstra, Jaco Hamman, Carrie Doehring, Pamela McCarroll, Susan Kassouf, Esther Huftless, Glenn McCullough, Bruce Rogers-Vaughn, Nancy Ramsay, Josh Morris, Rúben Aronja, and Lewis Rambo—good colleagues all and a balm for any voyage. I would not have embarked on this journey without the encouragement and guidance of Susannah Frearson—an editor for Routledge. Many thanks Susannah. I also deeply appreciate Mary Jeanne Schumacher and Cindy Geisen for their work and suggestions in reading each chapter. Naturally, any errors are mine. I am also grateful for Kathy McKay who read and critiqued the final draft. Kathy is an amazing copyeditor. My final note of indebtedness is to Cindy. She not only reads and critiques whatever writing project I undertake, she also adds helpful ideas and sources. In the vein of Huey Lewis and the News: Happy to be stuck with you. I'm so happy to be stuck with you.

Introduction

In a *Star Trek: The Next Generation* episode (season 6, episode 6), Amanda is a young woman assigned to the *Enterprise* to conduct research, which she loves. While on the ship, Amanda begins to have experiences that trouble her. Sometimes what she imagines in her mind happens in reality. Enter Q, a nearly omnipotent, eternal being. Q, annoyingly, has appeared to Captain Picard and his crew several times before. This time Q magically arrives on the *Enterprise* to deal with this young woman, who, unbeknownst to her, is a member of Q's people, if we can call them people. One of her parents was a human being, while the other was a Q who had escaped the community of Qs. I know, a lot of Qs. Of course, Amanda had no knowledge of this until Q arrived. At a crossroads, she can decide to remain human, losing all her Q-like powers, or choose to join Q and his people. Amanda is torn between remaining a limited human being and finishing her internship on the *Enterprise* or accepting becoming a being like Q.

At one point in the episode, while on the bridge with Captain Picard, Amanda listens to the distress of Lt. Commander Geordi La Forge and others. Geordi is on the surface of the planet the *Enterprise* is orbiting, trying to help engineers keep the planet from an impending climate catastrophe. The planet's dominant species have polluted their world for centuries, and the present generation, many of whom are sick, is desperately attempting to use science to prevent the total collapse of the global environment. As Amanda and the crew anxiously listen, it appears that climate collapse and extinction are imminent. At this point, Amanda decides to become a Q and she uses her powers to save the planet. The viewer sees the planet's atmosphere miraculously—more accurately, magically—cleansed and life restored. Who needs engineers and scientists to save a world when one has a Q? If only we had a Q.

While I love science fiction and magic, I recognize the fantasy of a Q saving us emerges out of a deep sense of powerlessness and anxiety in the face of present and future consequences of the climate crisis. Actually, the term "climate crisis" is a bit of a misnomer, primarily because it is not a singular crisis. Rather, it represents numerous intersecting and interrelated political, economic, scientific, and religious/spiritual crises. Western social imaginaries are

DOI: 10.4324/9781003518013-1

implicated in the rise to what we call the climate crisis—a polycrisis. Like the inhabitants of the planet, we are discovering that the Western social imaginaries we have constructed are part of the problem, not the solution. As Clayton Crockett (2012) argues, “We need to experiment radically with new ways of thinking and living, because the current [Western] paradigm is in a state of exhaustion, depletion, and death” (p. 165). The tragic reality is that these paradigms contribute not simply to the extinction of human beings, as a result of a massively degraded planet, but also the demise of millions of other species.

But what are these paradigms or social imaginaries? If we can identify these, can we, as Crockett hopes, not only critique them but create new theories that alter our ways of thinking and living?[1] This book sets out to do so with regard to psychoanalytic theory and practice. More specifically, I make use of the philosophy of Giorgio Agamben (and others—Rosa, Derrida, Latour, etc.) and Indigenous philosophies (and anthropologists) to critique as well as reimagine psychoanalytic theory and concepts. A question may arise as to why I use Indigenous perspectives. I find that philosophers such as Agamben are helpful in critiquing Western philosophical and religious anthropologies, but Indigenous narratives and practices can help us reimagine ways of thinking about and living with other species, as well as relating to the Earth.[2] It is necessary to stress that, like anthropologist Viveiros de Castro (2017), “we cannot think like Indians; at most, we can think with them” (p. 196). To think with them is not only an attempt to decolonize Western paradigms and epistemologies; it is also to learn how we might emend and use our social psychoanalytic imaginaries.

Naturally, other psychoanalytically informed therapists and theorists have taken note of the dire realities of climate change. Relying on the Psychoanalytic Electronic Publishing database, which currently has 152,000 articles and 141 books, one can obtain an overview of the interest in climate change among psychoanalytic writers. Between 2010 and 2019 there were 10 articles that mentioned “climate crisis,” while there were 167 since 2020. The phrase “climate change” comes up 3 times in the 1990s, 9 times in the 2000s, 187 times in the 2010s, and 443 times in the 2020s. The phrase “ecological crisis” has 88 mentions since the 1960s, with 55 of these since 2020. Clearly, these findings represent growing interest and anxiety about climate change within the past 15 years among psychoanalytically inclined scholars, though it is important to note these articles represent .0034% of the total number of articles in the database.

To draw closer to the literature will help illuminate the varied analytic perspectives and their foci regarding climate polycrisis, and, in doing this, I can situate this book within this literature, as well as indicate how it is similar to and different from these authors. Sally Weintrobe (2013, 2021, 2022, 2024) is a major contributor to the discourse on climate change and psychoanalysis.[3] She brings psychoanalytic concepts to address climate anxiety, neoliberal exceptionalism in creating a culture of carelessness, and the need to nourish an ecological self. Sally Weintrobe (2022) argues, “The problem for the lover of the natural world in a culture of uncare is how to maintain the ecological self" (p. 104). This is related to what Weintrobe calls "psychecide, meaning the overall demise of truthful reality-based and empathy-based thinking" (p. 98). "Sustaining (keeping stable and repaired) the ecological psychic imaginary," Weintrobe contends that this, "involves staying connected in a feeling-ful way with the natural world, with other living beings (human and non-human), and with the caring part of the self" (p. 109).

Weintrobe's colleague and, often, collaborator, Paul Hoggett (2010, 2012, 2013, 2019, 2023) addresses climate psychology in terms of difficult climate emotions, defenses, and perversions. Robert Samuels (2023), like Weintrobe, applies psychoanalytic theory to global issues, arguing that overcoming climate change requires moving beyond national politics and addressing unconscious fears about global governance. Renée Lertzman (2017) applies psychoanalytic theory to environmental psychology, arguing that many individuals experience a form of melancholia—a deep, unresolved grief over ecological destruction that manifests as apathy or disengagement rather than activism. Steffi Bednarek (2024), as editor, brings together a number of voices that, in critiquing (1) traditional psychotherapy for reinforcing unsustainable societal norms and (2) colonizing disciplinary regimes,[4] seek to advocate for a more holistic, intersectional approach to mental health in the face of climate change. Donna Orange (2017) uses psychoanalytic theory and concepts to understand the climate crisis and climate inaction, while also proposing a radical ethics for the climate crisis. Joseph Dodds (2012), relying on various psychoanalytic schools (e.g., Kleinian, Lacanian, Jungian, etc.) and the philosophy of Deleuze and Guattari, seeks to understand and depict the unconscious elements of the climate crisis. In a later work, Dodds (2019) contends that the psychological causes and consequences of the climate crisis are in need of psychoanalytic investigation. Using the work of Otto Fenichel, Dodds offers an eco-psychoanalytic approach to the climate crisis.

While there are others, these are some of the major contributors to the discourse on climate change and psychoanalysis. I mention these scholars to identify some of the themes and aims of their approaches, as well as what is missing. By and large, these authors demonstrate a deep and broad concern for what is taking place around the globe vis-à-vis climate change. They are therapists whose concerns are for patients, for human societies, and, thirdly, other species. They do not believe that the consulting room is to be separated from social, political, economic, and cultural realities, since these all shape and impact subjects' psychic lives. In other words, they affirm their responsibility for the well-being of their patients and all other human beings. Toward this end, they bring to bear psychoanalytic theories (and philosophies) and concepts to understand and respond to climate emotions, as well as sources of climate action and inaction. Relatedly, they use psychoanalytic theory to critique systemic realities, such as neoliberal capitalism, Western colonization, nationalism, and cultural anthropocentric imaginaries with the goal of offering possible ways forward. In other words, as psychoanalytically oriented therapists and cultural critics, they seek, in my view, to raise what is unconscious to consciousness in the hope that awareness offers a chance to change or to choose an action toward the real social, political, and economic sources of suffering.

I have found their interdisciplinary work not only needed and commendable but also enlightening. And yet, there are some lacunae. In general, the tendency is to avoid a critical analysis of psychoanalytic theory itself. How have psychoanalytic theories, constructed from the tools of Western philosophies, contributed to the polycrisis we face? Does psychoanalytic theory (or theories) need to change or can it change given the revelations of the climate polycrisis? Another general absence entails a lack of attention on other-than-human species and their sufferings/traumas. To be sure, many of these authors are concerned about the decline of insect and fish populations, as well as the increase in extinctions of other species. Nevertheless, the lack of attention to other species is, in my view, a symptom of the chasm between human animals and other species—a chasm produced by Western political philosophies and theologies, which I will elaborate in Chapter 1. One may respond to this general omission by noting that therapists are concerned about and know more about the psychology (and suffering/trauma) of human beings. That's fair, but, if psychoanalytic therapists are going to stray from their lane by analyzing social, political, and economic realities, why not consider the psychologies and sufferings of other species? Is there an unconscious belief that human psychologies are more important than the psyches, for lack of a better term, of other species? Are human psyches radically different, indicating an epistemological rift between human beings and other species? Is the omission another consequence of implicit Western anthropocentrism, exceptionalism, and exemptionalism embedded in our philosophies and psychologies? Are human beings psychologically affected by the sufferings and extinctions of other species? Consider Donna Orange's (2017) important book on radical ethics for the climate crisis. While she addresses anthropocentrism and exceptionalism, other species are mentioned primarily as an aside or afterthought. Orange writes, "I believe, actually, that only a radical ethics of the fundamental worth of every human life will make the difference we need in the climate crisis" (p. 120). I agree, in part. However, there is no mention of the fundamental worth of other species, which is a question of both ethics (Nussbaum, 2022; Singer, 1975) and psychology, as I hope to make clear in the coming chapters. Radical ethics and psychology include acknowledging and respecting the singularities of human beings and other species. This is an ethic that radically alters our perceptions, dispositions, and behaviors toward other species and the Earth.

This said, let me briefly summarize each chapter, which will illuminate the overall aims of this book. Chapter 1 depicts the ontological rift, its attributes, and its effects vis-à-vis the apparatuses of Western philosophical and scientific traditions. Since Freud relied on Western philosophies and views of science in the construction of his theories. this chapter explores and critiques Freud's views of nature, civilization, and Indigenous peoples. In addition, I highlight, from a psychoanalytic-ecological perspective, some of the consequences of the ontological rift embedded in Freud's theories. Because the ontological rift is pervasive in Western philosophical and scientific traditions, the chapter begins with a lengthy explanation of the ontological rift and a depiction of its features as they relate to how Western human beings construct and relate to "nature." From here I address some of the psychosocial dynamics and consequences of the rift. To illustrate the ontological rift and its consequences manifested in Freud's works, I then discuss his interrelated views of nature, civilization, and native peoples, which, in addition, I view as symptoms of the rift.

With the argument that Freud's psychoanalytic theories emerged out of Western philosophical and Cartesian–Baconian scientific beliefs, a question emerges regarding how a psychosocial developmental perspective can avoid the architectural flaws of Western thinking. How can we decolonize Western philosophy and science in developing a theory of psychosocial development? To provide *an* answer to this question is the focus of Chapter 2. More particularly, I argue in this chapter that good-enough parents' care for their infants represents a type of anarchic care. This anarchic care is foundational to infants' actualizing their potential for their initial semiotic resonant dwelling in the world without the rift. Yet, as (Western) children develop, they eventually internalize the dominant societal apparatuses of the rift, and the early form of resonant dwelling recedes, replaced by semiotic forms of dwelling informed by the ontological rift. I will depict this trajectory but, at the same time, argue for a developmental perspective that offers another possible viewpoint of embodied-relational, resonant dwelling that renders the rift inoperative, which we see glimpses of in Western life and in Indigenous ways of dwelling in the world. Since parents' care is foundational for infants' experiences of survival and thriving, I begin with a definition of care before delving into a discussion of what is meant by anarchic parental care. From here I focus on the psychosocial and embodied-eco-resonant relational development of infants and children, relying primarily on object relations and intersubjective theories, as well as philosophical perspectives—Western and Indigenous.

Given the climate polycrisis we face and the ontological rift between human beings and other species, and reimagining psychosocial development that spans the rift, how might we reimagine the notion of trauma, which is a central idea of any psychological theory? Chapter 3 addresses the notion of trauma as it relates to human beings *and* other species. If we are to cross the ontological divide, we need an idea of trauma that includes the sufferings of human beings and other species. To make my case, Chapter 3 begins with a discussion regarding trauma and its varied characteristics as it pertains to human beings. Once this is accomplished, I amend the notion of trauma with the aim of including other species, recognizing that the specifics of trauma will be unique to each species. For instance, the traumas of precocial asocial species, such as reptiles, and altricial social species, such as human beings and dogs, will differ considerably. I illustrate this emended version of trauma, relying on Jack London's story about Buck (a domestic dog), whose world is shattered after he is brutally abducted and shipped to the Alaskan territory. This will include identifying the varied sources of Buck's traumas, as well as the source of his recovery, namely anarchic care. The chapter ends with some implications of this perspective for psychoanalysis.

A key clinical concept in psychoanalysis is "transference," which has also proven to be helpful in understanding extra-clinical communications between human beings and human interactions with other species and the Earth. Indeed, the apparatuses of the ontological rift discussed in Chapter 1 represent a collective negative transference vis-à-vis other species. In Chapter 4, I claim that the concept of transference can be understood as a semiotic phenomenon, which pertains to other species and their relation to their local environment. More specifically, the chapter reframes *transference as a semiotic phenomenon of dwelling* and *eco-transference as animistic resonant dwelling*. In addition, this chapter explicates and differentiates between what I call anthropocentric transference and ecological transference.

Another key concept in psychoanalytic theory is "resistance." Chapter 5 starts by attending to the meanings, attributes, and limitations of the concept of resistance. From here I consider resistance in light of the ontological rift, semiosis, and its consequences, namely the disquiet dissonance of Western subjects—a form of defensive resistance against the reality of existential insignificance. This provides the foundation for expanding our understanding of resistance by returning to psychosocial development, animistic epistemologies, and the work of Giorgio Agamben. These are key for depicting ecological-political forms of resistance—forms that render the apparatuses of the ontological rift inoperative. All of this provides the basis for the final chapter on the cultivation of ungovernable animistic selves in the Anthropocene Age.

The crises associated with climate change, which seemed to give rise to authoritarians and the political, economic, and ecological crises they produce, raise questions about how to respond. Clearly, the emergence of authoritarians around the world is, in one sense, a response to the present and looming climate disasters and accompanying traumas. In terms of psychoanalytic theories and practices, how do we understand not only our role in relation to the needs, concerns, and sufferings of patients, but our role in terms of the larger society, as well as our ethical responsibilities and commitments to othered human beings, other species, and the Earth. In the final chapter, I explore the relation between psychoanalysis and the notion of ungovernable selves. I begin by discussing how analytic therapies emerged at a particular Western cultural moment that had previously been dominated by commitment therapies of various religious traditions. By juxtaposing these therapies, we gain clarity about the strengths and limitations of both but, more importantly, we begin to take note of *one* possible aim of the analytic process, namely, the facilitation of ungovernable selves who choose commitments toward the real sources of suffering—human and other-than-human. From here I move to address the notions of rebellion and revolution, using Albert Camus's work, primarily because it is important to make distinctions between these concepts and ungovernable selves. I then explicate the notion of ungovernable selves and its relation to psychoanalysis. The concluding section of the chapter shifts to the notions of "soul" and "spirit." I do this for two reasons. First, psychoanalysis has, from its inception, had an ambivalent relation to religions (including so-called primitive Indigenous spiritualities), often failing to acknowledge the affinities between religion and science. Second, one way to understand the notions "soul" and "spirit" is in terms of abstraction and resonance. That is, they are, at best, abstractions for experiences of resonant relations of belonging, which are features of ungovernable selves. In this case, psychoanalytic practice can consider patients' religious beliefs and experiences in light of whether or not they are connected to ungovernable selves' resonant experiences and relations with other human beings, other species, and the Earth.

Before embarking, I would like to offer several clarifications. First, the scientific literature on climate change is vast and readily available. In the last three decades, a cottage industry developed trying to coin an apt appellation for the global climate crisis. Crutzen and colleagues (2000), at the beginning of this century, called our current era the Anthropocene, arguing that human beings are responsible for climate change. Jason Moore (2016) believes that term is too broad, failing to clearly depict the true origin of our crisis. He uses the term "Capitalocene," because capitalism and the attending apparatuses of market societies fetishize profits and relentlessly exploit human beings, other species, and the Earth. As Lynne Segal (2023) notes, 1% of the population—the polluting elite—produces more carbon emissions than the bottom 10% (p. 165). Capitalism, which produces massive wealth and income inequalities (e.g., classism), in its extractive and exploitative fetishized aim of profit is, according to Moore, responsible for both the climate crisis and resistance to climate actions. David Runciman (2023) similarly believes the term "Anthropocene" falls short, largely because other human beings and societies (e.g., Indigenous, poor countries, and poor people within imperialistic nations) are not responsible. Like Moore, he notes that capitalism and its apparatuses are largely liable for greenhouse gas emissions. However, we cannot understand the emergence of capitalism in the West without noting the rise and prevalence of nation-states, especially the Western colonizing leviathans (Britain, Spain, Portugal, United States, etc.) that enslaved and exploited peoples around the world. "Leviathan" is a reference to the political philosophy of Thomas Hobbes, which argues that sovereignty and its attending apparatuses are essential for a stable ordering of society (nation-state) and the exercise of individual freedom. The leviathan (state and its apparatuses) exercises the use or threat of political violence to enforce the social contract and maintain societal security and stability. Runciman contends that the Western notions of sovereignty and nation-states, entwined with capitalism and imperialism, are to blame for the current climate crisis. Whatever we call it, climate polycrisis is here and will continue to deepen in the decades ahead, negatively impacting human beings and other species.

Of course, it is important to stress that not everyone is impacted equally. Wealthy nations and wealthy people (top 10%) will have more resources, as well as political and economic power, to secure resources that 90% of the world lack. Indeed, we have already seen billionaires prepare for the disasters ahead by building secure bunkers or mega-yachts for family and friends (Meister, 2024). The movie *Elysium* is a fictional story where the wealthy have created an orbiting system where they escape the harsh conditions of a degraded Earth, where the residents are exploited by capitalist leaders. While fictional, it is based on present-day realities. There are, then, the realities of classism, racism, and other forms of exploitation, oppression, and marginalization in the market societies of the world, which will only worsen as we sink deeper into the polycrisis. As an aside, the suffering of other species or their presence on the orbiting station is absent from the film—in my view, evidence of unconscious speciesism.

Another point to make is that the psychological sciences, as noted above, are focused almost exclusively on human beings. While this seems to be an obvious choice, there are a couple of issues to be noted. First, the term "psyche" in philosophy and psychology is used almost always in reference to human beings and not other species. It is not simply that this represents a kind of privileging of human beings, it also is a symptom of what is discussed in the next chapter, namely the ontological rift between human beings and other species. To be sure, there are important distinctions to be made between the psyches of human beings and those of other species, but "distinction" does not necessarily mean separation. Second and relatedly, the emergence of quantum theory and scientific and philosophic discourses on questions of the relation between the "mind" (and consciousness) and material reality has revealed how interconnected the lives of innumerable species and material existence are (see Barad, 2007; Chopra et al., 2017; Hart, 2024; Penrose et al., 2017). An implication is that human psyches, while different, are part of the web of biodiverse "psychic" life, which suggests a kinship that many of us tend to avoid. Let me add here that Indigenous philosophies, in general, hold to the premise that all living beings possess a psyche, whether that is understood as soul or semiotic agency. Moreover, as Dana Lloyd (2024) writes, the Yurok believe "the Land itself is a living being. According to Yurok earth-based theology, 'the Earth is alive and has a spirit'" (p. 121). This contrasts with Western epistemologies that treat the Earth as an object—dumb, mute, and, therefore, exploitable without care or remorse.

Given this recognition, this book is focused on psychoanalytic theory vis-à-vis human beings. Nevertheless, this focus in no way is meant to privilege human beings over other species. Moreover, my aim is to offer a way of thinking (theory) that crosses the divide between human beings and other species. I hold, then, to the underlying premise that human minds are inextricably part of the psyches of other creatures—our kin. This premise has implications for theories of psychosocial development, as well as implications regarding an ethic of care and justice that include other species and the Earth. In other words, a psychological theory that crosses the divide or chasm between human beings and other species implicitly aims toward facilitating (1) the ethical dispositions of empathy and compassion toward other-than-human beings and (2) resonant belongingness with these beings and the Earth.

These aims are not simply and solely aims of an eco-philosophy or eco-psychology. That is, I believe psychoanalytic theories, implicitly or explicitly, are not only concerned with understanding and responding to the psychological suffering of persons in the consulting room and those in the larger society. Theories also focus on what a good life is. What do human survival and flourishing mean? The crises associated with climate change reveal that our survival and flourishing are contingent on the survival and flourishing of other-than-human animals, as well as insects, trees, and so on. We all inhabit and belong to this one Earth. This fact undergirds the categorical command to care for a biodiverse Earth.

As for this one Earth, many readers will no doubt be familiar with the scientific quest to find life and intelligence on other orbs. This quest is not merely out of curiosity. It is also fueled by the question, are we alone in the universe? This question can only emerge from dissonant subjects who are captive to apparatuses of the rift, which tells us that we humans believe ourselves to be the only intelligent species on Earth. If intelligence is believed to reside only with human beings, then, of course, we feel alone in the vast cosmos that we hope includes other intelligent beings among putative dumb animals. This is not only a tragic question; it is one that is absolutely absurd. While Indigenous peoples have long accepted the fact of multiple intelligences, those of us for the rift are only now beginning to realize that we are surrounded by innumerable intelligences that we have only begun to note, understand, and appreciate (Schlanger, 2024; Yong, 2023). Indeed, some of these intelligences will be here long after human beings become extinct. So, we are definitely not alone. We have numerous kin who we have yet to meet and understand. Life and intelligent life exist here, and therefore we are challenged to care for all of it, since we are all dependent on this one Earth.

Finally, psychoanalytic theorists and therapists, like most human beings, tend to be adept at seeing the speck in the eyes of others (theories) and not the beam in our own eyes. This book is an attempt to recognize this beam and to find ways to remove it so that we can see more clearly how our psyches are interrelated to the varied psyches of our other-than-human kin. This is my hope, but I also expect readers to take note of the beams or specks in my eyes and offer their own correctives for the sake of becoming aware of our perceptions, dispositions, and behaviors toward other species and the Earth.

As much as I would love to have a Q rescue us (present and future human beings and other species), I know it is a fantasy. Like all fantasies, nothing happens, though for a moment I am distracted from my helplessness and anxiety. Choosing an action targeting the real sources of the climate crisis will not take away my anxiety and helplessness, but the choice, commitment, and subsequent action are better alternatives than fantasies. My actions and the actions of others will not likely result in enough change to stem the coming climate catastrophes and their attending sufferings. Of course, our actions might, but my ecological choices, commitments, and actions are contingent not on hope but rather on the categorical command to care. As Barry Lopez writes, "Successfully locating the proper frame of mind and then acting is not, I think, about refusing to accommodate fear. It's about the cultivation of love" (2022, p. 200). Valerie Kaur (2020) adds: "This is our defiance—to practice love even in hopelessness" (p. 241).

Notes

1 Philosopher Lewis Gordon (2023) points out that the word critique comes from "the ancient Greek verb *krinein* (to decide), from which emerged not only the nouns *kritēs* (judge) and *kritērion* (means or standard of judgment) but also *krisis* (crisis). A crisis is a point at which a decision must be made. … Critique, then, is also a form of crisis" (p. 154). The climate polycrisis demands critique, especially of those apparatuses that are, implicitly or explicitly, implicated in the emergence of this crisis, which, I argue in this book, are Western sciences, including, in part, the psychological sciences.

2 There is disagreement about whether to use lowercase or uppercase when referring to this planet. I use uppercase to signal the distinctiveness of the Earth, as well as to give a nod to Indigenous philosophies that view the Earth in terms of sacredness or spirit. My aim is not to idealize, but rather to point to a kind of disposition and behavior toward an object one conceptualizes as a living entity.

3 Linda Aspey, Catherine Jackson, and Diane Parker (2023) edited a work on reviving psychological and spiritual agency in the face of the dire realities of the climate crisis. Sally Weintrobe contributed the foreword. I mention this work as a related example of using psychology to understand and respond to the climate crisis.

4 See also Celia Brickman's (2018) and Sally Swartz's (2018) arguments that the origins of psychoanalysis are founded on Western philosophies that produce not only the ontological rift between human beings and other species, but also Western racist and gendered representations of colonized peoples. They seek to decolonize psychoanalysis (see also LaMothe, 2023).

Chapter 1

Freud, Nature, and Civilization

Western Philosophical Apparatuses and the Production of the Ontological Rift in Psychoanalysis

We are funny animals—not in the sense of being humorous, though we certainly can be, but rather in the sense of being a bundle of contradictions. We exhibit incredible feats of intellectual acumen while also acting ignorantly and stupidly. Human beings can be altruistic, generous, and kind while also being horribly selfish, sadistic, and brutal. We can be astonishingly creative at being terribly destructive. We are conscious of who we are while often blind to the consequences of what we are doing. Indeed, we can, like Freud, create complex methodologies for becoming more aware while, at the same time, these very methods of cultivating awareness keep aspects of our lives in the shadows. Human life, it appears, is a never-ending affair of navigating these contradictions by critiquing what we believe to be true. This ongoing process of examining ourselves and exposing our contradictions, Socrates believed, makes life worth living, individually and collectively, though his life ended as a result of his method, suggesting that critical, honest examination threatens those who possess power and privilege. If we situate Socrates in the context of the Athenian agora, then the aim of intersubjective critical analysis or self-examination is, in my view, about learning to dwell with each other and the world more justly and more caringly, even in the midst of our manifold contradictions. Put differently, for Socrates, self-examination is a political obligation that exposes illusory, cherished premises and shallow arguments, which partially explains his demise. Many human beings wish to remain in the depths of the cave. The tragic contradiction is that many human beings express a desire for self-knowledge and for seeking the truth, but, when the truth is brought to our attention, we deny, undermine, resist, or destroy it.

Since the time when Socrates proclaimed both the importance of self-examination and his own lack of wisdom, Western philosophies, theologies, and sciences[1] have produced elaborate systems of thought or what Cornelius Castoriadis (1997) called "social imaginaries." These social imaginaries, whose destructive premises remain largely uncriticized, have been instrumental in justifying slavery, the exploitation of workers, and the colonization of other peoples and lands (Chibber, 2013; Mills, 1997, 2017; Robinson, 1983). Yet, in the 20th century the spirit of Socrates would bloom in a

DOI: 10.4324/9781003518013-2

number of fields. Edward Saïd (1979) wrote a signature work wherein he argued that Western literature constructed "Orientals" in decidedly negative ways. Similarly, Stuart Hall (1997, 2016) highlighted the distorted Western representations of colonized persons and their cultures. A parallel shift was taking place in cultural anthropology. Anthropologists such as Eduardo Kohn (2013), Eduardo Viveiros de Castro (2017), and Tim Ingold (2013, 2022) set aside Western prejudices regarding Indigenous cultures and began to understand the complex epistemologies of native peoples. In the mid-20th century, there were also glimmers of a critical examination or decolonialization of psychiatry and psychoanalysis with the emergence of Frantz Fanon (1952/2008), though this would not bear fruit until the early decades of this century (see Gibson & Beneduce, 2017; Parker & Pavón-Cuéllar, 2021; Sheehi & Sheehi, 2022; Sheehi, 2018; Swartz, 2018).

These important critiques entail the recognition that human science theories and concepts have, wittingly and unwittingly, served as apparatuses that have produced distorted representations of other human beings, which has accompanied profound and destructive failures in understanding and relating to othered peoples. I suggest that the crises of European colonizing states during and after World Wars I and II served as an impetus not only for revolutions of independence among colonized peoples but also for the burgeoning critical literature of postcolonial and decolonial approaches in science, philosophy, theology, and the arts. Indeed, Julian Go (2016) notes that the third wave of postcolonial discourses created a flurry of researchers who are interrogating their theories and practices: decolonizing feminism (e.g., McLaren, 2017), social theory (e.g., Go, 2016), education (e.g., Battiste, 2013), anthropology (e.g., Harrison, 1997), psychoanalysis (e.g., Gherovici & Christian, 2019; Simek, 2011), and botany (e.g., Schlanger, 2024). These approaches aim to identify, critique, expose, and excise Western representations of and misguided premises about marginalized and oppressed peoples, whether they live or lived in or outside the metropole. Justice for and the flourishing of colonized peoples were the aims of these critical stances that emerged as a result of European nation-state crises.

Another crisis, one that is global and dire, offers opportunities to expand critical decolonizing methodologies toward examining our relations to other species, plants, and the Earth (see De Vos, 2023; Meijer, 2019, 2020; Nussbaum, 2022; Schlanger, 2024; Singer, 1975; Yong, 2023). The climate polycrisis, in other words, reveals how Western philosophies,[2] theologies, sciences, and economics operate as apparatuses that produce representations of other species and the Earth that function, consciously or unconsciously, to legitimate the brutal and destructive instrumental uses of other-than-human species and the Earth while, at the same time, ignoring, dismissing, or denying their suffering. Given the realities of the climate crisis, we in the West are discovering that our cherished and often unquestioned anthropocentric anthropologies and reifying sciences comprise models and premises that distort "nature,"[3] other species, and our relations with them. As noted in the introduction, Clayton Crockett (2012) suggests, "We need to experiment radically with new ways of thinking and living, because the current paradigm is in a state of exhaustion, depletion, and death" (p. 165). New ways of thinking and dwelling require a critical stance toward the narratives, theories, rituals, and so on, that inform our work and our ways of being in the world.

Before we experiment radically with any tradition, we first need to examine critically its foundations. In this chapter, I depict the ontological rift and its effects related to the apparatuses of Western philosophical and scientific traditions. I do so because Freud's views of nature, civilization, and Indigenous peoples emerge out of these traditions. In addition, I highlight, from a psychoanalytic-ecological perspective, some of the consequences of the ontological rift embedded in Freud's views. Because the ontological rift is pervasive in Western philosophical and scientific traditions, I begin with a lengthy explanation of the ontological rift and a depiction of its features as they relate to how Western human beings construct "nature." I next address some of the psychosocial dynamics and consequences of the rift. Then, to illustrate the ontological rift and its consequences manifested in Freud's works, I discuss his interrelated views of nature, civilization, and native peoples, which I view as symptoms of the rift.

A number of important clarifications must be made before embarking. First, because the ontological rift is central to my argument regarding its connection to psychoanalysis and to Freud in particular, I spend considerable time and attention explicating its attributes. This provides a hermeneutical framework for briefly addressing Freud's views on nature, civilization, and Indigenous peoples. Second, anyone familiar with psychoanalysis knows that, while Freud founded psychoanalysis, psychoanalytic theories and research abound. The result is that much has changed in the last century. So, why begin with Freud? One reason is that Freud remains influential. Another related reason is that much of psychoanalytic theory, even with important and necessary changes, continues to maintain some of the same philosophical and scientific premises that undergird the ontological rift.[4] For instance, Harold Searles (1960) and recently Susan Kassouf (2017) note the absence of theorizing regarding the psychosocial development of other species compared with that of humans. In my view, this lacuna is a consequence of the ontological rift. Another lacuna is the poverty of literature regarding the notions of trauma, consciousness, and agency in other species (Coburn, 2024; LaMothe, 2023). Let me add Katie Gentile's (2018) comment as further evidence of the rift: "Psychoanalysis usually unreflectively describes eating animals as an indication of psychological health, whereas vegetarianism is still often seen as an indication of an eating disorder or repression of oral aggression" (p. 7). In a later article, Gentile (2021) notes that vegetarianism and animal rights have been of little interest to psychoanalysts, though when vegetarianism is addressed it is not in a positive light. These examples are, to my mind, illustrations of the ongoing ontological rift in psychoanalytic theorizing, all of which began with Freud.

Of course, some may counter by contending that psychoanalysis is focused on the psyches and communications of human animals rather than other species. This focus, they might argue, would explain the lacunae. Fair enough, but this would overlook a fundamental fact of human psyches' relational connection not simply to other human beings but to the environment, which necessarily includes other species. Consider numerous Indigenous narratives that are populated by the psyches of other animals (see Erdoes & Ortiz, 1984; Kerven, 2018). In other words, the premise, in my view, of Indigenous narratives is that one cannot understand the human mind without exploring its connections to nature, which includes other species. This is analogous to Winnicott's (1965) comment that "there is no such thing as a baby" (p. 39). Winnicott wanted to study parent–infant interactions, as well as these interactions situated within the social-political environment. The climate crisis reveals that human psychic well-being is contingent on a biodiverse Earth, which indicates that our psyches are connected to the environment, which, of course, includes other species. I add here that important distinctions are to be made between the minds of human beings and other species, but differentiation or distinctions do not necessarily imply separation or a rift. The human psyche, in brief, is not radically different or separate from the semiotics of other species (Hart, 2024). Anthropologist Eduardo Kohn (2013), like Charles Sanders Peirce (Peirce, 1991) a century ago, contends that "all life is semiotic and all semiosis is alive" (p. 117).

A third clarification concerns my intention not to bash Freud or to suggest that his and subsequent psychoanalytic theories are fatally flawed. Indeed, taking a page from Freud (1910/1957a),

> When research … approaches one who is among the greatest of the human race, it is not doing so for the reasons ascribed to it by laymen. "To blacken the radiant and drag the sublime into dust" is no part of its purpose.
>
> (p. 63)

There are, instead, two purposes for critically engaging Freud. One purpose of criticism is revision. Another aim is heuristic, inviting readers to critique the theories and traditions they rely on to engage or dwell in the world.

A fourth and related clarification concerns Freud and philosophy. Freud, like us, was a person of his time. He consciously and unconsciously internalized Western philosophical traditions as well as adhered to Enlightenment values of progress, autonomy, reason, and science, though with important qualifications (Breger, 2000; Hanly, 2020; Sulloway, 1992). The Western philosophies that informed Freud's theories and views of science contained and produced the ontological rift, but let me rush to add, this is not to suggest that these philosophies or Freud's theories are without merit, meaning, or value. Rather, critique, as earlier noted, is for the sake of amending. As a self-proclaimed scientist, Freud, I believe, would be open to correction and revision—philosophical or otherwise.

Since I have mentioned Western philosophical traditions, let me point to Freud's later strong ambivalence toward philosophy. Freud was familiar with and mentioned numerous Western philosophers, from Plato and Aristotle to Schopenhauer and Schleiermacher (see Gay, 1988; Ricoeur, 1970; Sulloway, 1992). Indeed, Freud's core idea of psychosocial developmental is based on a Greek myth, which in turn is, as I will explain below, ensconced in an essentialist or ontological Greek (and Israelite) political philosophy, namely, patriarchal sovereignty. Yet, he later was dismissive of philosophy, even as he used it. Freud (1933/1964c) wrote that philosophy

> behaves like science and works in part by the same methods; it departs from it, however, by clinging to the illusion of being able to present a picture of the universe which is without gaps and is coherent, though one which is bound to collapse with every fresh advance in our knowledge.
> (p. 160)

A few pages later, Freud added, "You will scarcely be able to reject a judgment that philosophy of today has retained some essential features of animistic mode of thought" (p. 165). This was not a positive observation, and his negative view of philosophy is evident in a letter to Josef Popper-Lynkeus in which Freud (1932/1964b) noted that, in the beginning of his career, "The psychology which ruled at that time in academic schools of philosophy had very little to offer and nothing at all for our purposes" (p. 219).[5] Freud believed in pursuing truth and science, and science's methods were, he believed, qualitatively better than philosophy. He nevertheless relied on Western philosophies and their anthropological premises.

Like Freud, I rely on the works of Western philosophers,[6] though I take pains to highlight what philosopher Giorgio Agamben (1999) calls an "architectural flaw" in Western philosophies, theologies, and sciences. Highlighting a flaw, however, does not suggest being dismissive of Western philosophy. Instead, it implies a call for revision. This revision relies on the use of various philosophers in conversation with present-day anthropologists, who have long recognized Western hermeneutical biases that obstruct understanding Indigenous epistemologies that were and are foundational to their survival and thriving.[7] These Indigenous epistemologies can point the way toward emending the architectural flaw, which is the focus of this book. A final clarification: this chapter is a heuristic, not an exhaustive, examination of the ontological rift in Freud's views on nature, civilization, and Indigenous peoples. The next chapter offers a developmental perspective that builds on this critique.

The Ontological Rift in Western Philosophies and Sciences

Agamben (1999) writes that a burning house reveals "the fundamental architectural problem [that] becomes visible for the first time" (p. 115). We might imagine the burning house as the climate crisis. For Agamben, the architectural problem is the "deep ontological rift … between animal and human" (Dickinson, 2015, p. 173). That is, the rift entails "a radical and total discontinuity between human and nonhuman" (Kompridis, 2020, p. 252), which is a part of the assumptive world of most Western persons.[8] Agamben (2004) writes:

> It is as if determining the border between human and animal were not just one question among many discussed by philosophers and theologians, scientists and politicians, but rather a fundamental metaphysico-political operation in which alone something like "man" can be decided upon and produced. If animal life and human life could be superimposed perfectly, then neither man nor animal—and, perhaps, not even the divine—would any longer be thinkable.
>
> (p. 92)

Bruno Latour (1993) similarly notes that there are, in the West,[9] "entirely distinct ontological zones: that of human beings on the one hand; that of nonhuman on the other" (pp. 10–11). Likewise, Jacques Derrida (2008) referred to this as an "abyssal rupture" (p. 30) between human beings and other species.[10] Agamben, Latour, and Derrida contend that Western philosophies, theologies, and sciences serve as apparatuses that produce and maintain this abyssal rupture between human animals, other animal species, and plant life (see Schlanger, 2024). Add to this Isabelle Stengers's (2023) contention that a central apparatus producing and exacerbating this rift is global capitalism, which fosters a kind of stupidity or lack of critical thinking pervasively and perversely evident wherever capitalism has created market societies.

Interestingly and ironically, there is clear evidence that Freud (1933/1964d) recognized this rift long before Agamben, Derrida, Latour, and Stengers. Freud wrote that human beings "have no business to exclude themselves" from the whole animal kingdom (p. 204), which is an idea going back to Aristotle and brought to further light by the works of Charles Darwin—works that Freud would have been familiar with. In another work, Freud (1917/1955a) claimed that the emergence of civilization accompanied the human's

> dominating position over his fellow creatures in the animal kingdom. Not content with his supremacy, however, he began to place a gulf between his nature and theirs. He denied the possession of reason to them and to

> himself he attributed an immortal soul, and made claims to a divine descent which permitted him to break the bond of community between himself and the animal kingdom.
>
> (p. 140)[11]

Freud added, but did not pursue the implications, that "this piece of arrogance is still foreign to children, just as it is to primitive and primaeval man" (p. 140).[12] In his last work, Freud (1939/1964a) believed he was "diminishing the gulf" between humankind and other species (p. 100). Freud, remarkably and to his credit, put his finger on a central problem in Western political philosophy, theology, and Cartesian–Baconian science.[13] Moreover, I believe Freud (1917/1955c) wanted to expose the narcissistic and arrogant illusions human beings have regarding their position in the cosmos and in relation to other species. Admirably, he hoped to diminish this gulf but, as I will argue below, he did not succeed—at least not enough—largely because he continued to rely on the foundational epistemological premises associated with Western philosophy as well as Western political philosophies' unquestionable adherence to the view that civilization and sovereignty are necessary for human survival, freedom, and political life.[14] In short, Freud's insights regarding the rift did not serve as a critical hermeneutical framework in constructing his theory of human development. This means that his theories unwittingly served and serve as apparatuses for producing and maintaining the rift. Not surprisingly, Freud's insight, hope, and unconscious replication of the rift is another instance of the contradictory nature of human beings.

Before expounding further on the attributes of the rift, it is important to mention briefly that this rift can exist between human beings. Cultural anthropologists Eduardo Kohn (2013), Eduardo Viveiros de Castro (2017), and Tim Ingold (2013, 2022) point to the great divide in Western engagements with Indigenous peoples, which I will say more about below. For now, I note that Western political philosophies, theologies, and literature have served as apparatuses for constructing native peoples as inscrutable, inferior, ignorant, childlike, savage, or primitive (Hall, 1997, 2016; Saïd, 1979, 1994), which legitimate and justify racist, sexist, and classist forms of oppression, marginalization, and enslavement by so-called reasonable, civilized peoples (see Baptist, 2014; Danner, 2009; Desmond, 2023; Mills, 1997, 2017; Patterson, 1982; Wilkerson, 2020). Consider a relatively benign and common illustration of this rift in the midst of caring for othered people. In the late 19th century, Frank Linderman befriended Plenty Coups and the Crow people. Indeed, he chronicled Plenty Coups's life. Yet, Linderman included a telling remark regarding his experiences of living with the Crow people:

> I am convinced that no white man has ever thoroughly known the Indian, and such a work as this must suffer because of the widely different views of life held by the two races. … I have studied the Indian for more than

> forty years, not coldly, but with sympathy; yet even now I do not feel that I know much about him.
>
> (Lear, 2006, pp. 1–2)

I think Linderman, to his credit, recognized the gap between his knowing (and way of being in the world) and that of the Crow people. He, of course, did not recognize that the beam was in his own eye—a beam of Western philosophical premises that undermined his ability to understand the Crow people. An additional example of this interhuman rift: philosophical and human science apparatuses have often constructed a sizable rift between "sane" human beings and those who are constructed as mad (Gilman, 1985a, 1991; Porter, 1991a, 1991b). The history of the treatment of people suffering from mental illness in the West is disturbing, despite its benefits (Szasz, 1977). It is disturbing, in part, because of how often these persons were treated as objects to be studied and controlled.

We can gain a better picture of the rift by identifying and discussing its attributes and premises. The main attribute of the rift, which is evident in Western political philosophies and the attendant colonization of othered peoples, is depersonalization. This can be understood by first explicating what is meant by personalization or personal knowing. Simply stated, personalization involves the recognition and treatment of an individual as a unique, valued, inviolable, and responsive, agentic subject (see Løgstrup, 1997; Macmurray, 1961). For philosopher John Macmurray (1961), echoing Kant's categorical imperative, personal knowing entails recognition of the individual's singularity,[15] which, in terms of social-political life, means that the notion of personhood ideally acts as the philosophical and political basis for the ideas of justice, ethics, and politics. I will say more in the next chapter, but, for now, personalization or personal knowing is also the foundation for acts of care and subsequent experiences of trust that are necessary for the survival and flourishing of children (and adults). This is not just the case for psychosocial development; it is also the case for a viable politics—civic care and civic trust and fidelity. To be a person is to be included in the polis and to possess, actually or potentially (e.g., infants), a political-social agency. Also, to be a person is to be due justice.

This view merits a qualification. In Western political philosophies, recognition of other human beings as persons has varied and still varies in degree. In Aristotle's polis (see Aristotle, 1971), women were included in the polis and deemed to be persons, but they were denied political agency. They belonged in the polis, but they were, at best, second-class citizens. As second-class citizens, Athenian women were socially-politically constructed as possessing a diminished capacity for reason when compared with adult male citizens.[16] This became the justification for restricting women's participation in many areas of social-political life, which continued for millennia in the West and exists in various patriarchal societies today. Also, in Aristotle's polis, barbarians and slaves were seen as human beings but not quite full persons in that they too were excluded from political life. In my view, any social-political construction that falls short of recognizing an individual's singularity is considered a form of depersonalization. This suggests that depersonalization can be assessed on a continuum, with one end being absolute denial of personhood (e.g., enslavement, state-sanctioned torture, killing, ethnic cleansing—social death [Patterson, 1982] or bare life [Agamben, 1998]), and at the other end is the partial denial of personhood (e.g., denying women the vote). The former represents an ontological abyss in the sense that it is seemingly unbridgeable, while the latter is a rift that may or may not be bridged.

The idea of personalization in Western political philosophies excludes other animals.[17] Aristotle affirmed that human beings are political animals. As political animals, human beings exercise political agency and freedom,[18] while other species, since they are not persons, are believed to lack both.[19] This means that other species (and the Earth) have been rendered politically irrelevant, except in cases of their use value. In other words, some species, of course, are included in the polis. They are used for agriculture, war, construction, pleasure, and other human activities. They nevertheless are constructed as lacking personhood or singularity and agency, which justifies or legitimates their instrumental use and exploitation. Put differently, because other species are perceived to lack personhood, they exist outside the zone of ethics and justice. Given the ontological rift, the ideas of politics and justice do not apply to other species, though there have been notable exceptions in the philosophical tradition regarding this (see footnote 3 in this chapter).

Some may point out that other species do indeed lack political agency and capacities for reason, which would seem to logically exclude them from political consideration. I have three interrelated responses to this. First, infants have only a nascent or potential social-political agency and reason, but we nevertheless consider them to be an integral part of the polis. Second, other species are an integral part of the biosphere, and the *very existence of the polis is dependent on a robust and diverse biosphere.* As Terry Eagleton (2016) notes, a biodiverse Earth "is the first condition of our existence" (p. 228). This implies that any notion of the political or any notion of the human being would necessarily be inclusive, recognizing the singularities of all species. This would not simply diminish the rift; it would deny its existence (see Meijer, 2019, 2020; J.-J. Rousseau, 2016). Third, other species share with human beings existential realities of precarity, vulnerability, and resilience. The polis, at its best, constructs apparatuses that facilitate resiliency, reciprocity, and cooperation for the sake of survival and flourishing. What the climate crisis reveals is that fostering the resiliencies of and cooperation with other species not only benefits the polis; these qualities are required for maintaining a biodiverse Earth. To recognize precarity and vulnerability and to foster resiliency depend on personal epistemologies—recognizing the singularities of other human beings and other-than-human species.

Two other features of personalization are worth mentioning because of their relation to depersonalization and the ontological rift. Personalization (and depersonalization) is a species of semiosis, which, in turn, includes the capacity for valuation (see Faber, 2023, pp. 128–129, 308–309; Morris, 1964, p. 15). To recognize the singularity of another individual is to affirm their existential value as a person—as an inviolable, responsive subject—and this involves knowing, which is based in semiosis. For Macmurray (1961), personal knowing, to be ethical, must place forms of semiotic instrumental or object knowing in the background. Imagine good-enough parents whose caring interactions are grounded in the recognition of their children as persons. This personal recognition and knowing have an attendant valuation, which is that children are valued in themselves—and thus *a categorical respect* for children's subjectivity follows. Any object or instrumental knowing (use value) in parental care is secondary to personal knowing. For instance, a good-enough parent's analytic assessment of and attunement to an infant's distress call entail temporarily placing object knowing in the foreground and personal knowing in a supporting role. For Macmurray, this is ethical because it is for the sake of the infant's well-being and it is momentary. In this case, these are not two separate ways of knowing. They are integrated. That said, depersonalization represents the splitting-off of personal knowing from object knowing, which is a feature of the ontological rift.

The capacity for valuation in human beings is complicated because valuation can be connected to beliefs in inferiority and superiority—a form of semiosis. When these evaluative beliefs are present, one finds forms or gradations of depersonalization and attendant instrumental or object knowing, which produces a rift between the knower and the object. For Aristotle, while human beings are political animals, other species are inferior because they lack human reason and agency. He understood women to have "inferior" capacities for reason and agency and placed enslaved individuals even lower on the scale. Beliefs in superiority and inferiority are responsible for legitimating and justifying all kinds and gradations of depersonalization, such as the one Aristotle proposed.

Whether constructing those outside the group as inferior or those within the polis as persons but inferior (women), the result is an ontological rift and an existentially insecure subject. This dependence on inferior others, in other words, is fundamentally and perpetually insecure and precarious. Members of the "superior" group must continually construct apparatuses to affirm their belief in their own superiority by ensuring that othered human beings and other species experience being inferior or adopt the role of inferiority. Put another way, the group uses the belief in their superiority to affirm their existential or ontological significance while denying or minimizing the significance of those constructed as inferior. As Massimo Leone (2020) notes,

> Just as the instinct for survival prevents human beings from fully coming to terms with mortality, so too the semiotic instinct impedes them to becoming aware of that which is, nevertheless, an absolutely central element of both individual and collective life: insignificance.
>
> (p. 8)

The reality of existential insignificance elides beliefs in superiority or inferiority. By contrast, depersonalization's ontological rift is produced when beliefs in, or more accurately illusions of, superiority and inferiority are used in constructing social-political life.

Before saying more about existential insignificance, I need to address several features of personalization, depersonalization, and the ontological rift. Let me begin positively. Personal knowing is an epistemology that represents a disposition, methodology, and virtues. The disposition is openness or receptiveness, perhaps even respect and curiosity, vis-à-vis the other. The symbol "person" signifies the other as a valued, inviolable, and responsive subject.[20] This symbol is, at first, empty in the sense of waiting for the other to give concreteness to the appellation—to appear as itself and for itself. To recognize the other as a person does not and cannot exhaust the meanings associated with the singularity of the individual. A good-enough parent who has just given birth recognizes the infant as a person, and this recognition accompanies the disposition to receive the infant's communications—to be open to the infant appearing as they are.

This disposition implies what I call an epistemological methodology. By this I mean that personal knowing entails an approach to learning about the other. In other words, personalization creates a relation (Buber's [1958] I–thou, if you will) or a space between the knower and the object such that the object can reveal itself, become itself, to actualize its potentialities.[21] Personalization is unlike instrumental epistemologies (I–it relations) that seek to learn from the object only those things necessary for "it" to be of use. Personal epistemologies are unlike Cartesian–Baconian scientific and technological epistemologies that aim to know data about the object (Macmurray's object knowing), such as whether that object is a person, another species, or a material object. I will return to this, but for the moment let me say that cultural anthropologists have noted that Indigenous peoples' personal knowing extends to other species and material objects (see Ingold, 2013, 2022; Kohn, 2013; Viveiros de Castro, 2017). As Viveiros de Castro (2017) notes, indigenous epistemologies are personal in the sense that the "form of the Other is the person" (p. 61). The aim of personification is that "one must know how to personify because one must personify in order to know" (p. 62). Viveiros de Castro continues by stating, "All animals and cosmic constituents are intensively and virtually persons, because all of them, no matter which, can reveal themselves to be (transform into) a person. This is not a simple logical possibility but an ontological potentiality" (p. 57). He adds, "Personhood and perspectiveness—the capacity to occupy a point of view—is a question of degree, context and position rather than a property distinct to specific species" (pp. 57–58).

I add here that these anthropologists are not saying that an Indigenous person who recognizes a crow as a person believes that the crow is a human. The person is not anthropomorphizing the crow. It is a crow-person, possessing its own singularity. Indigenous personal knowing represents an epistemological approach to creating a relation, a space, for the other species to appear as they are—learning about other species and the world—an approach that is essential to Indigenous people's survival and thriving. This said, I am not suggesting Indigenous peoples do not "use" other species in a form of instrumental knowing. I am saying that personal knowing is the primary method of learning, while instrumental knowing is secondary. This suggests that Indigenous people, unlike subjects of the rift, include other species in their politics and ethics.

The third feature of personal knowing concerns the virtues of attention and respect. Parents who recognize their children as persons are attentive to their communications, and this attentiveness accompanies respecting children's ways of being in the world. An illustration is helpful here, both in regard to personalization and depersonalization. Ruby Sales, an African American activist, grew up in the Jim Crow South, where oppression and violence impacted African American adults and children. Sales recalls her childhood living in the midst of the white racial contract:[22]

> I grew up in the heart of Southern apartheid. And I'm not saying that I didn't realize that it existed, but our parents were spiritual geniuses who created a world and a language where the notion that I was inadequate or inferior or less-than never touched my consciousness. I grew up believing that I was a first-class human being and a first-class person, and our parents were spiritual geniuses who were able to shape a counterculture of black folk religion that raised us from disposability to being essential players in society.
>
> (Tippett, 2016)

Her parents, in the midst of the depersonalizing apparatuses of white supremacy, were consistently attentive to their children, which clearly accompanied innumerable actions that communicated respect. This provided the space for Ruby and her siblings to experience themselves as persons—unique, valued, inviolable, responsive subjects.

What does all this mean with regard to depersonalization and the ontological rift? Depersonalization establishes a relation and disposition whereby the other is denied singularity. Depersonalization's epistemological approach is to obtain information about the object only in terms of its use value or, in the worst cases, to simply deny that the object has any value at all (e.g., ethnic cleansings). If the object, which could be human, is deemed to have no use value, it is discarded, destroyed, or ignored. If there is any respect, it is only for maintaining the object's use value. That is, attention and respect are conditional. For instance, a farmer raises cattle and cares for them with the aim of selling them to be slaughtered for their meat and other products. The farmer necessarily knows a great deal about cattle, but this knowledge is devoid of personal knowing. It is, instead, instrumental and technological, which manifests an ontological rift between the farmer and the cattle. To personalize the cattle would be counterproductive to the capitalistic enterprise and its instrumental epistemologies because it would comprise learning about and respecting each steer's and cow's singularity. Put differently, the presence of the rift and depersonalization places the cattle in a zone of non-justice, which means that any discourse of legitimation or justification has no bearing on the cattle's existence except with regard to their use value. Personalization, by contrast, means the cattle fall within the zone of justice and ethics (Nussbaum, 2022; Singer, 1975), which then complicates the farmer's relation to the cattle.

There is a final related feature of depersonalization and the ontological rift. In the discussion above regarding beliefs in superiority and inferiority, as well as epistemology, I limited, for the moment, any elaboration of personalization and depersonalization as signifying operations or semiosis. The ontological rift, in other words, is produced by apparatuses that are semiotic in nature. This topic will be discussed in later chapters. The term *significance* derives from the Latin *significantia*, which denotes meaning or energy. The infinitive *signficare* denotes to mean, signify, or import. Both the Latin and the English refer to what Charles Sanders Peirce called semiosis—the theory of signs. Semiosis is more comprehensive than language or linguistics, which was a focus of Ferdinand de Saussure. All of this is quite complex and unnecessary to delve into here. For my purposes, semiosis includes all or any signifying operations, whether they are signs (e.g., a bee dance), symbols, or language, and so on. These signifying operations, according to Charles Morris, entail four uses, namely, informative, valuative, incitive, and systemic. Morris (1964) continues:

> Signs may be used to inform someone of the properties of objects or situations, or to induce in someone preferential behavior toward some objects or situations, or to incite a specific course of action, or to organize the dispositions to behavior produced by other signs. … [I]n general, designative signs are used informatively, appraisive signs are used valuatively, prescriptive signs are used incitively, and formative signs are used systemically.
>
> (p. 15)

Morris is referring primarily to human beings, but he is not restricting signs to language, symbols, rituals, and so on. Rather, any sign, for Morris, has these uses, whether it is associated with bees or bears. In terms of personalization, signifying refers to valuation whereby the other is valued in itself, in its singularity. Depersonalization of the ontological rift is a form of semiosis whereby the other is denied significance or the object's existential existence is reduced to mere use value, like the farmer's cattle. Depersonalization, in other words, is a semiotic operation that denies, consciously or unconsciously, the singular existential significance of the other. Indeed, depersonalization can be said to assign existential significance to oneself and existential insignificance to depersonalized others.

When speaking about depersonalization of the ontological rift as a form of semiosis, it is necessary to mention the issue of power. Foucault (1972) recognized that knowledge(s), which is based on semiosis, and power intersect. The notion of power is complex, contested, and cannot be addressed here. Instead, I contend that depersonalization, which entails a type of knowing (and, paradoxically, a refusal to know), also manifests in a type of relational or political power that forces the object to conform to projections, which I will say more about in Chapter 4 regarding transference. To force an object to conform to projections signifies power *over* the object—a kind of dominion or domination that is dependent on semiotic beliefs in superiority and inferiority—and, consequently, a denial of its singularity. Put differently, depersonalization requires the power to control and subjugate an object, which is the very foundation of Cartesian–Baconian science as well as most Western philosophies. Immanuel Kant (1980), for instance, believed that human beings had no moral duties to other animals, so other-than-humans could, therefore, be treated as a means to our ends. The categorical imperative does not apply to other species, which is a premise that justifies domination. John Stuart Mill, echoing Bacon, argued that human beings must conquer nature (Eberhart, 2024, pp. 140–141). Domination and objectification or depersonalization are evident in the techno-scientific control of nature and other species, which has been observed by critical theorists such as Adorno and Horkheimer (Rigby, 2017, p. 278). The farmer's knowledge of the cattle, for instance, accompanies the power (e.g., technological, scientific, political, economic) necessary to impose his will on them to obtain their use value. The farmer is also manifesting a refusal to know in the sense of being open to knowing the singularity of each animal. The farmer's use of power over the cattle, in other words, reflects a willed ignorance and lack of curiosity regarding individual steers and cows. For the cattle, this is a negative form of power. To shift to human beings, Aristotle's partial depersonalization of women requires the exercise of political and economic power (dependent on varied apparatuses) to force and coerce women to live out the expectations of men in power. Orlando Patterson's (1982) notion of social death in terms of slavery, Agamben's (1998, p. 83) discussion of bare life, and Fanon's (1952/2008) "zone of nonbeing" (p. xii) represent types of depersonalization that entail brutal objectification of and power over other human beings (and, I would add, other species).

I contrast this kind of negative power with the positive power of personal knowing (acceptance of ignorance, which initiates curiosity) that creates a relation or space for others to appear as they are and to actualize their potentialities instead of demanding them to conform to our wishes and desires. Depersonalization is the exercise of negative power in the sense of collapsing this space. It may entail forcing the individual to realize their use value or excluding them, whether that is moving them from habitable spaces (e.g., removal of Indigenous people from their lands, removal of other species) or through death (e.g., ethnic cleansing, industrial slaughter factories).

The apparatuses of negative power and depersonalization produce and are evidence of the ontological rift, which can also be grasped in terms of the notions of potentiality and actuality, both of which are important in addressing the psychosocial development of human beings. The notion of potentiality stems from Aristotle's work regarding the relation between potentiality (*dynamis*) and actualization (*energeia*). For Agamben, Western philosophical traditions have largely

> subordinated potentiality to actuality: so we begin with the actual, speaking humans and their political and artistic productions, and we see potentiality at present as a capacity or skill that is defined by the final action. We see potentiality as secondary or accidental.
>
> (Colebrook & Maxwell, 2016, p. 188)

This is derived, in part, from Aristotle's notion that "actuality is prior to potentiality" (as cited in Ugilt, 2014, p. 26), though this does not mean that Aristotle believed that "potentiality exists only in actuality" (Agamben, 1999, p. 180). What is important here is that the potentiality of an individual being is actualized by its coming into existence through, in the case of social animals, nurturance and protection. Depersonalization, which is a form of instrumental, objectifying knowledge and power as dominion, denies, in whole or in part, the individual's actualization of potentialities. Racism, sexism, ableism, classism, imperialism, and speciesism represent types and gradations of depersonalization that thwart or deny the actualization of potentialities.

In summary, the ontological rift can be depicted in terms of depersonalization, the features of which become visible when depicting personalization. Personal knowing entails recognizing the other as a unique, valued, inviolable, responsive subject—a singularity. The sign "person" is an empty sign in the sense that it does not exhaust the meaning and value of the individual. Personal recognition is the foundation for affirming togetherness or belongingness and therefore entails the inclusion of the individual being in discourses regarding politics, ethics, and justice. By contrast, depersonalization denies (or diminishes) the other qua person, which leads to the exclusion of the individual from the realms of noninstrumental care, justice, and politics. There is, in other words, a rift between those deemed to be persons and depersonalized individual humans (or other species), which is manifested in exclusion from political and social belonging. I also argued that personal knowing is an epistemology that represents a disposition (receptivity), methodology (learning from and about the other), and virtues (respect and attention). All of this creates a relational space for the other to appear—to actualize their potentialities. Depersonalization, by contrast, possesses the disposition of closedness and an epistemological methodology of willed ignorance and instrumentality, as well as inattention and disrespect. This forecloses the relational space, inhibiting or obstructing the depersonalized other's appearances. I also addressed personalization as a species of semiosis that involves the valuation of individuals in themselves. This valuation creates a relational space for the other to appear in their singularity. Depersonalization is a form of semiosis that assigns insignificance to other species or significance merely in terms of the object's use value. Finally, personalization represents a type of knowing and power that creates spaces for others to actualize their potentialities—an absence of projection and the presence of ignorance or curiosity. Depersonalization, by contrast, represents a form of instrumental, hierarchical, and objectifying knowing and negative power as dominion (control over) that forces others to submit to projections, which undermines or inhibits their ability to actualize their potentialities.

Psychosocial Sources, Dynamics, and Consequences of the Ontological Rift in Western Philosophies and Sciences

Questions remain. Why was the ontological rift created, despite its negative consequences? What are some of its psychosocial functions, dynamics, and consequences? The above discussion partially answers these questions. For instance, in terms of psychosocial functions, beliefs in the superiority of human beings over other species provide some sense of shared esteem or significance and power for human beings by projecting onto others existential insignificance. But more remains to be said in terms of the psychosocial functions of the rift. I have also already briefly alluded to some of the consequences of the ontological rift, such as (1) the exclusion of othered human beings and other species from the political realm (zones of non-justice), (2) the instrumental and objectifying use of othered human beings and other species, and (3) the insignificance of othered human beings and other species represented by the neglect of their needs and experiences. Here I turn to a brief consideration of the psychosocial functions, dynamics, and consequences of the rift. I rely on psychoanalytic concepts, which are important in the discussion of Freud's views regarding nature, civilization, and Indigenous peoples.

Let me begin by contending that the apparatuses of the ontological rift are semiotic collective delusions,[23] though they have very real and damaging effects, as evident in the climate crisis and the sixth extinction event (Klein, 2014; Kolbert, 2014). In other words, the ontological rift is not an existential reality or fact but rather a Western human semiotic construction that serves to provide individual and collective esteem and sense of control, all the while avoiding, by way of rationalization, moralization, and denial, the stark reality of existential insignificance. Above I mentioned Leone's (2020) contention that our instinct for survival prevents us from coming to terms with our mortality, and our semiotic instinct impedes us from the central reality of all life, which is our existential insignificance. The fact of our existential insignificance evokes, consciously and unconsciously, fear and anxiety about the precarity of our significations (Strenger, 2011, pp. 4, 35).[24] The semiotic construction of the ontological rift, in my view, functions to assert human significance while projecting existential insignificance onto othered human beings, other species, and the Earth. This represses anxiety associated with existential insignificance. In other words, Western philosophies and theologies ontologize and eternalize human significance. Signs (abstractions) such as God, Plato's Ideal forms, immortality, eternal life, the kingdom of God, and others are signifiers not only of the rift but also of the denial of our anxiety and fear associated with the realities of existential insignificance. When Freud (1915/1957b) wrote, "In the unconscious everyone is convinced of our immortality" (p. 296),[25] he was, from my lens, pointing out that (Western) human beings fear, resist, and deny not simply the reality of death but also our existential insignificance.

There is another clue in Freud's work regarding self-esteem that relates to the ontological rift. Recall that Freud acknowledged that (Western) human beings have created a rift between themselves and other animals. What I find interesting is that Freud (1917/1955a) viewed this as a sign of collective arrogance, which he said is absent in children (p. 140). This is a very telling and important observation, one that I will return to later. For now, let me say that arrogance connotes superiority and dominion, which are features of adults' self-esteem. Apparently, children's senses of self-esteem and control are not dependent on the belief in human superiority or on having power over other species (or othered human beings).[26] Yet, eventually children of the rift will, through processes of internalization, come to adopt this arrogance and its attendant beliefs.

Arrogance can be further understood in terms of the psychoanalytic notion of projection, which I mentioned briefly above. The ontological rift represents the projection of inferiority onto other species, which affirms the contingent, arrogant belief in human superiority. Restated, those holding on to the apparatuses of the ontological rift project onto other species existential insignificance while ontologizing human significance via signs such as kingdom of God, paradise, eternity. This projection allows persons of the rift to deny or repress their existential insignificance, to moralize their instrumental use of other species, and to feel no remorse for the suffering of other species (e.g., in industrial slaughter factories) and maintain indifference regarding the destruction of other species and their habitats.

As noted above, it is not simply other species that have suffered as a result of the ontological rift. Philosopher José Ortega y Gasset (1957) wrote that human beings live "in perpetual fear of being dehumanized" (p. 27). Apparatuses that produce acts of dehumanization or depersonalization[27] represent the ontological rift among human beings. Terms such as social death, bare life, and zone of nonbeing express the reality of insignificance forced, violently and in other ways, onto other human beings. The fear of being dehumanized is connected not to an indifferent universe (contra Strenger, 2011) but to the reality that other human beings can willingly depersonalize other human beings. Depersonalizing other human beings is inherent in the ontological rift's delusions.

Implicit in this discussion is the idea that psychosocial projections (which depend on socially constructed apparatuses) must be always operative in order to shore up and maintain the esteem and dominion that are dependent on inferior others. In other words, we are not conscious of the rift or of its delusions because the apparatuses of the rift are pervasively (and perversely) operating, fostering a sense of the truth or reality of these delusions. This provides a sense of security and control vis-à-vis our individual and collective identity and attendant self-esteem. But I would argue that what remains unconscious is fear and anxiety concerning our existential insignificance. That is, those of us of the ontological rift live lives of quiet (and not so quiet) desperation. This desperation becomes evident when the apparatuses are under threat and our fragile identity and its dependence on insignificant others are exposed. Illustrations are seen in white rage and hatred whenever people of color assert their rights—their significance (see Alexander, 2010; Anderson, 2016). It took over seven decades for women to obtain the right to vote in the United States. Leaders of this movement faced harassment and death threats from those whose identity was dependent on the subordination and putative (political) insignificance or inferiority of women. In terms of climate change, I suspect that much of the pushback by climate deniers is related to their unconscious fear and anxiety over losing their cherished identities as (eternally) superior to and dominant over other species.

Speaking of the unconscious, in psychoanalytic parlance, I have argued that those of us who live out of the ontological rift split off insignificance and impermanence and project these onto "inferior" other species. These psychosocial dynamics can be further understood in terms of varied notions of the unconscious. Lynn Layton's (2020) notion of the "normative unconscious," for instance, is applicable here. Human beings are deemed to be normative, which is evident in Western social imaginaries that are anthropocentric. Other species, with regard to anthropocentric social imaginaries, fall into the category of nonnormative and are relegated to the unconscious. Put another way, since other species are considered insignificant in themselves (except for their utility), they cannot be considered by human beings to be normative. Naturally, othered human beings are nonnormative in relation to human beings who believe themselves to be superior. In either case, the normative unconscious means that nonnormative others do not make it into our consciousness unless they possess instrumental significance (like the farmer's cattle). Another way of saying this is that we simply do not think about nonnormative others.

The normative unconscious has implications for narrative constructions with regard to both the present and the past. Today, we simply do not construct our narratives, by and large, in terms of other species. There are occasional references or stories about other animals, but the *norm* when constructing social-political narratives is the human.[28] In addition to this, by making human beings normative, the intelligences of other species are excluded or, if recognized, deemed to be inferior.[29] Closely associated with the normative unconscious is Timothy Zeddies's (2002) idea of the historical unconscious. Zeddies's notion refers to how some (othered) human beings are excluded from history and its narratives. Women in patriarchal societies and religions are rarely central figures in historical narratives. In the United States, black women and men are largely written out of dominant historical narratives because of systemic racism that constructs them as insignificant (nonnormative) or significant only in terms of bolstering white peoples' illusions of their own superiority and dominance. In terms of other species, the historical unconscious clearly applies. If some species make it into the history books, it is to advance the human story. In other words, the singularities of other species are absent. For instance, in the Exodus myth, God sends a fifth plague that decimates horses, donkeys, camels, herds, and flocks (Exod 9:1–4), which is yet another illustration of instrumental epistemologies and the disavowal of the singularities, significance, and dignity of other species. In this story, they function anonymously in terms of their use value with regard to coercing the Pharoah. Moreover, the species mentioned in this and other stories are taken as a group. Their individual significance is denied.

I add to this by suggesting that there are a political unconscious and a developmental unconscious. Andrew Samuels (1993, 2001, 2004) and others (Fanon, 1952/2008; Laubscher et al., 2022; Layton et al., 2006) have consistently sought to understand human subjectivity in light of the political. In terms of human beings, those who are denied political agency and significance are relegated to the political unconscious or what Fanon called the zone of nonbeing. In patriarchal political societies, men are regarded as normative, and women are largely excluded from political (and historical) consciousness. In terms of the ontological rift in Western political philosophies and theologies, other species are denied both political agency and importance. Through being denied political significance, other species, by and large, exist in the zone of non-justice—the ideas of justice and the political simply have no relevance to them.[30]

Earlier, I mentioned Howard Searles's (1960) observation that psychoanalytic developmental theories overlook other species. "It would seem then," he suggested, "that a natural next phase would consist in our broadening our focus still further, to include the investigation of man's relationships with his nonhuman environment" (p. 23). I propose that the reason for this lacuna is due to the ontological rift.[31] Of course, other animals are mentioned in terms of psychosocial development,[32] but I would argue that Searles's observation continues to be true, as Susan Kassouf (2017) argues. And I suggest that this falls under the category of the ontological rift and its attendant developmental unconscious.

The four types of the unconscious described above and their relation to the ontological rift can be further illuminated in terms of Jean-Paul Sartre's (1965) notion of bad faith, Herbert Fingarette's (1969/2000) view of self-deception, and Donnel Stern's (1997) concept of weak dissociation. For Sartre (1965), bad faith "has in appearance the structure of falsehood. Only what changes everything is the fact that in bad faith it is from myself that I am hiding the truth" (p. 150).[33] Sartre also noted that "bad faith" does not come from the outside to human reality. "One does not undergo his bad faith; one is not infected with it; it is not a state" (p. 150). Sartre used the singular, but the same can be said of the plural. Families, communities, and societies can hide truths from themselves, which is, I suggest, the reality of existential insignificance screened by social, political, and scientific apparatuses that produce the rift. Bad faith or hiding truths is not quite fair. A better term is self-deception. Philosopher Herbert Fingarette (1969/2000) asked, how is it that someone or a people actually deceive themselves? "It is," he wrote,

> when we judge that there is a purposeful discrepancy between the way the individual really is engaged in the world and the story he tells himself that we have the complex but common form of self-deception in which we are interested.
>
> (p. 62)

This self-deception results from not spelling out one's engagement in the world or, better, from spelling out one's engagement such that one does not avow one's actions or the consequences of these actions. A very common example is going to the grocery store and buying meat, deceiving ourselves about the massive cruelties done to other species so that we can eat meat without remorse. We tell ourselves stories that spell out our ways of dwelling in the world such that the sufferings and deaths of other species are elided.

Relying on Fingarette's work, psychoanalyst Donnel Stern (1997) argued that some individual and collective forms of not spelling out are illustrations of weak dissociation. Weak dissociation involves narrative rigidity, which is a complex form of organizing one's experience such that one's actions and consequences are narrowly spelled out. Inflexible narration, in other words, means that ideas, meanings, values, and affects that are unconsciously perceived to challenge the dominant story are excluded or unformulated. In weak dissociation, Stern argued, we spell out only what

> we believe we can tolerate, or that further our purpose, or that promise a feeling of safety, satisfaction, and the good things in life; we dissociate the meanings that we believe we will not be able to tolerate, that frighten us and seem to threaten the fulfillment of our deepest intentions.
>
> (p. 128)

As a result of "so insistently turning our attention elsewhere … we never even notice alternative understandings. Focal attention under these conditions is controlled by the intention to enforce narrative rigidity" (p. 132). If one wishes to test out this rigidity, just ask people to give up meat and to be more empathic to those species selected for human consumption. I recall a philosopher whose friends told her to read a recent book that, they argued, would motivate her to become a vegetarian. When asked whether she had read it, the philosopher laughed and confessed that she had not. The reason she gave was that she did not want to give up eating meat. This is not so much self-deception as it is weak dissociation and a concomitant refusal to be moved by stories that would interrupt her beliefs and accompanying privileges, which are embedded in rigid narratives that rationalize and moralize eating meat. The philosopher preferred to live out of the rigid narratives of the rift.

The terms bad faith, self-deception, and weak dissociation are relevant when it comes to understanding the ontological rift and its social imaginaries. Western philosophies, theologies, and sciences have, in part, functioned to spell out our engagement in the world such that we overlook how our actions have contributed to the vast destruction of other species and the Earth, which the climate emergency is making evident. The climate crisis, in other words, reveals the ontological rift and its accompanying types of the unconscious, bad faith, self-deception, and weak dissociation, which are not benign but catastrophically destructive. Indeed, relegating other species and othered human beings to the unconscious only serves to maintain our destructive relations while retaining our fragile Western illusions of human existential significance and our dominion or power over other species.[34] This also reveals the tragic reality of Western subjectivities captive to the apparatuses of the ontological rift. I do not think that human beings, as a species, are tragic creatures in themselves, any more than other animals are tragic. Although we are contradictory, we are not necessarily tragic. I do not, for instance, see the inherent tragic nature of Indigenous peoples who are able to live in reciprocal, cooperative, and sustainable relationships with other species and the environment. By contrast, Western human beings, generally speaking, *are* tragic because, by creating a rift and ontologizing our significance and dominion at the expense of the significance of the singularities and agencies of other species, we inadvertently undermine our (and Indigenous peoples' and other species') survival and flourishing. In short, our Western social imaginaries, which human beings create, represent bad faith, self-deception, or weak dissociation. As Charles Morris (1964) points out, "Man lives in his symbols, but he often maims himself, and even dies, because of them" (p. 87). This seems to be the case for Western subjects who are captive to the symbols of the ontological rift.

There is one more feature to address before moving to Freud. The ontological rift establishes a perennial or interminable contentious, conflictual relationship with that which is othered, whether it is nature, othered human beings, or other species. Consider, for example, Carlo Strenger's (2011) phrase "ontological protest of subjectivity" to refer to the putative "universal" tendency to seek control in the face of existential helplessness and the so-called indifferent universe.[35] I would agree with this formulation if it referred to Western human beings who construct and are captive to the ontological rift. The ontological rift produces subjectivities that inherently protest their existential helplessness and insignificance. Put another way, the term *protest* reveals a *contentious and conflicted subjectivity and relationship*, in this case to a universe perceived to be indifferent, or to nature, which is deemed to be cruel. In terms of other human beings who are constructed as insignificant, there is only self-deception, enmity, resentment, and bitterness in these relationships, where those who consider themselves superior must always be on alert for any protest or any assertion of the "inferior" other's significance. In our relations to other species, they are to be controlled, managed, tamed, subjugated, exploited, and so on, which is evidence of the exercise of negative power. Any "protest" (cries) of other species is largely ignored by virtue of our collective self-deceptions and weak dissociation. In other words, cooperation, reciprocity, respect, and concern/care are absent or nearly absent in relationships preoccupied by conflict, control, and negative power over those who are constructed as existentially insignificant—human and/or other-than-human species. Western subjects, captive to the rift and living lives of quiet desperation in the face of existential insignificance, remain in permanent conflict with whatever is on the other side of the rift, which is evident in Freud's conceptions of human beings and nature.

In summary, the ontological rift comprises shared delusions (e.g., human superiority over the inferiority of other species and othered human beings) that provide individual and collective senses of self-esteem and control or dominion in relation to those who have been othered—depersonalized. These delusions accompany psychosocial defenses of rationalization, moralization, splitting, and denial in relation to depersonalized beings. Second and relatedly, the shared psychosocial operation of projection entails locating existential insignificance in "inferior" species while securing significance for human beings deemed "superior." All of this results in perennial unconscious insecurity, anxiety, and fear that function to motivate human beings to construct and maintain apparatuses that secure human significance at the expense of other species. Third, I argued that the notion of the "unconscious" can be framed or understood in four ways, namely, in terms of the normative, historical, political, and developmental unconscious, which have varied negative consequences for othered beings. I further depicted notions of the unconscious in terms of bad faith, self-deception, and weak dissociation, which operate to rationalize and moralize our destructive ways of dwelling with other species and the Earth. Finally, the ontological rift produces subjects that are in interminably conflictual relations with whatever is believed to be on the other side of the rift.

Freud and the Ontological Rift: Nature, Civilization, and Indigenous Peoples

The categories of "nature," "civilization," and "Indigenous peoples" intersect in Freud's work. To discuss one category necessarily entails the others. Together, they are illustrative of the whole cloth of the architectural flaw of Western epistemologies and anthropologies, though, as we will see, Freud seemed to have some ambivalence about the use of these terms. These categories represent the epistemological premises and "legal" justifications not only for the colonization of other peoples and lands but also for the exploitation of other species and the Earth. As I noted in the introduction, psychoanalysts today may have largely shed these views of nature, civilization, and primitive humans or savages, but this does not mean the ontological rift has been abandoned in psychoanalytic theorizing.

Nature

Freud's views of nature are contradictory. He wrote, "When man personifies the forces of nature he is again following an infantile model" (1927/1961a, p. 22). Continuing, Freud remarked that "it is in fact natural to man to personify everything he wants to understand in order to later control it (psychical mastering for physical mastering)" (p. 22). Leaving aside, for the moment, issues associated with the deeply problematic notion of personifying nature and its relation to control or mastery, it is intriguing that Freud the scientist, the man of reason, proposed a few pages earlier a personifying Hobbesian view of nature—a view that justifies the existence of civilization. If civilization ceased to exist, he wrote, "She [nature] would destroy us—coldly, cruelly, relentlessly" (p. 15). Nature's cruelty and indifference, which are clear personifications, serve as the reasons human beings "came together and created civilization, which is also, among other things, intended to make communal life possible" (p. 15).[36] Indeed, "The principal task of civilization … is to defend us against nature" (p. 15).

This defense is limited because, Freud (1927/1961a) argued,

> No one is under the illusion that nature has already been vanquished; and few dare hope that she will ever be entirely subjected to man. There are elements, which seem to mock human control: the earth, which quakes

> and is torn apart and buries human life and its works; water, which deluges and drowns everything in a turmoil; storms, which blow everything before them.
>
> (pp. 15–16)

These examples reveal, for Freud, that the "forces of nature [rise] up against us, majestic, cruel and inexorable; she brings to mind once more our weakness and helplessness which we thought to escape through the work of civilization" (p. 16). This gendered view of nature is evident in an earlier work on a psychoanalytic study of Leonardo da Vinci. Freud (1910/1957a) opined that Leonardo was the "first man since the time of the Greeks to *probe* the secrets of nature while relying solely on [instrumental] observation and his own judgments" (p. 122, emphasis mine). Freud added, "If we translate scientific abstraction back again into concrete individual experience, we see that the 'ancients' and authority simply correspond to his father, and nature once more becomes the tender and kindly mother who had nourished him" (p. 122). Whether cruel or kind, nature is gendered. This represents an anthropomorphizing tendency in Freud, which he regarded as a feature of childhood or of "primitive peoples," certainly not something a person of science would do—except perhaps rhetorically.

While this gendering of nature begs for analysis, I highlight here a few key features with regard to nature and the ontological rift. First, Freud's epistemology regarding nature represents two types of knowing. On the one hand, he recognizes that human beings can personify nature[37] and other species with, for Freud, the aims of controlling the object to meet human needs and desires and perhaps to deflect our attention away from our existential helplessness and weakness. This form of personal knowing is part of all human beings and indicates the human capacity to recognize "beings who at bottom are like himself" (1927/1961a, p. 22). Jessica Benjamin's (1995) view of identification as likeness in difference and difference in likeness fits well with this kind of knowing. Personal knowing allows us to see the other as like us yet different. On the other hand, Freud the scientist is following a Western view of the relation between human beings and nature, going back at least to the science of Aristotle. Aristotle, long viewed as the first scientist, considered human beings to be animals—part of nature—though significantly different. As a scientist, Aristotle set up elaborate hierarchal taxonomies and dissected (probed, depersonalized, objectified) other species to learn about them.

Following but emending Aristotle, Francis Bacon (1561–1626), a devout Anglican philosopher and scientist, is credited with proposing the "inductive-experimental method as a replacement for Aristotle's methods" (Polkinghorne, 1988, p. 16). Bacon was not alone in making significant changes to science and, like others, he believed that "the practical aim of improving humanity's lot [depended on] the increased understanding and *control* of nature" (Grayling, 2019, p. 197). What Aristotle and Bacon have in common is a form of knowing that is instrumental and objectifying—depersonalizing, as noted above, in relation to the object being studied. Put differently, this form of knowing collapses the dialectical tension between likeness in difference and difference in likeness, leaving other species without singularity—simply and merely objects. If other species, as part of nature, are seen, they are nevertheless mere objects serving human needs and desires. While Freud *may* have personified nature as a rhetorical device, he was nonetheless echoing science's (and philosophy's) objectification and instrumentalization of nature (including other species) for the sake of knowledge, control, and protection.

Hartmut Rosa (2020) writes that "the normalization and naturalization of our aggressive relationship to the world is the result of a social formation … that is based on the structural principle of dynamic stabilization and on the cultural principle of relentlessly expanding humanity's reach" (p. 8). Like Freud, "We are structurally compelled (from without) and culturally driven (from within) to turn the world into a point of aggression. It appears to us as something to be known, exploited, attained, appropriated, mastered, and controlled" (p. 14). It is not just science, as Freud understood it, but the cultural, political, and economic apparatuses that produce the rift wherein our relationship to nature is one of objectification and aggression. Carl Schmitt (2005) believed—wrongly, in my view—that the friend–enemy relation defined politics, but it is more accurate that this defines Western subjects' antagonistic relation and disposition toward nature.

There is another point regarding the presence of two epistemologies in relation to nature and other species. Cartesian–Baconian science, which Freud uncritically adopted, can provide an immense amount of knowledge, as Freud recognized and many of us acknowledge. There have been all kinds of scientific advances and achievements in Freud's era and our own. That said, the objectification or depersonalization of other species and the Earth results in the absence of the epistemological principles of perspectivism and multinaturalism. In explicating epistemological perspectivism and multinaturalism, anthropologist Eduardo Viveiros de Castro (2017) contends that "virtually all peoples of the New World share a conception of the world as composed of a multiplicity of points of view. Every existent is a center of intentionality apprehending other existents according to their respective characteristics and powers" (p. 55). He states further that "multiplicity can be taken as a kind of plurality [and] Amazonian multinaturalism affirms not so much a variety of natures as the naturalness of variation—variation as nature" (p. 74). Perspectivism and multinaturalism represent an epistemological stance of learning about other species *as they are* in their manifold singular natures and not as we would want them to be. Science's objectification or reification and instrumentalization of "nature" and the concomitant denial of personal knowing of other species mean that scientific knowing and personal knowing are split off,[38] resulting in the denial of the singularities of other species. In categorizing other species into taxonomic groups, scientific epistemologies exclude the epistemological principles of multinaturalism and perspectivism, which limits our ability to recognize and learn about the singularities of other living beings who dwell in and on the Earth.

While Freud did propose a personified idea of nature as kind and nurturing, I argue that his dominant stance was that nature is dangerous. Human beings, he claimed, developed civilization to defend against nature, which means that human relations to nature are characterized by interminable antagonism and protest. Freud (1930/1961b) wrote about "this subjugation of the forces of nature, which is the fulfillment of a longing that goes back thousands of years" (p. 88). Indeed, Freud (p. 92) extolled human achievements of violently extracting wealth from nature, which evidences a pervasive instrumental, objectifying, conflictual relationship with nature. Climate change reveals just how devastating this kind of thinking and relating is. The point here is that, on the one hand, there is a personification (presence of personal epistemology) of nature and, on the other hand, there are instrumental objectifying epistemologies of domination and control. These two forms of knowing are not integrated in Freud's work.

Of course, Freud (1933/1964d, pp. 204–205) believed, with Aristotle, that human beings are animals, which suggests that human beings are also a part of nature. Behind this conceptualization is a history in Western philosophical thought of distinguishing between two natures of human beings (see Honneth, 2023; Khurana, 2023). This is a complex history, but for my purposes it resides in the tendency of Western "civilized" human beings to differentiate between the "nature" of human beings as animals and their second nature (e.g., culture, freedom) that marks human beings as not just different from other species but radically different—the rift. This is evident in Freud's anthropology. For instance, he wrote, "We recognize as *cultural* all activities and resources which are useful to men for making the earth serviceable to them, for protecting them against the violence of the *forces of nature*" (1930/1961b, p. 90, emphasis mine). These two natures reflect conflict with the forces of nature and an internal psychic conflict in the sense of not fully accepting our being animals in the face of our existential insignificance.

Naturally, all human beings (indeed, all semiotic beings) have the capacity to differentiate between this and that object. But Western philosophy, which is evident in Freud's conceptualization of the two natures, differentiates such that there is *separation* between the object and the human being. Culture is associated with human beings. By contrast, "nature" is portrayed as cruel, indifferent, and relentless.[39] This is hardly a view of nature that invites cooperation, reciprocity, and so on. Just as we use our capacity to differentiate between ourselves and our enemies, we establish a relation of separation where the dialectical tension between likeness in difference and difference in likeness collapses—intersubjectively and subjectively. In Freud's anthropology, civilization (and culture) and nature are forever at odds, and this distinction or differentiation leads to separation. Separation is, I contend, part of the philosophical tradition of conceptualizing and differentiating the two natures of human beings.

Let me add that the construct of the two natures, or Freud's view of nature and the nature of human beings, is not only inextricably tied to the ontological rift; it also represents an epistemological mistake. Freud tends to construct "nature" as an entity that really exists, as if it exists in reality. Nature or Mother Nature is simply and solely a sign (a metaphor) that points to the human experience of trying to make sense of and relate to the larger world (environment) and to communicate this to others. Nature does not exist as a particular existent entity with which we are in interminable conflict. Nature, in other words, is an illusion, an imaginary idea, an abstraction, even though it is believed to be an actual, identifiable object. "Nature," in short, is a metaphor masquerading as a fact. This is analogous to Freud's view of religion and its construction and use of the concept of "God." To Freud, the concept of God is illusory and is connected, in part, to human beings' anxiety associated with their existential helplessness. To be sure, many human beings report having relationships with and experiences of God. The semiotic construct or abstraction is believed to be real, even if unseen. Nature and God are abstract, ethereal constructs in the sense that they are not concepts linked to actual identifiable objects in reality (such as a dog, cat, etc.). Nevertheless, these abstractions reveal the kinds of dispositions and relations persons have in the world.

Epistemological illusions and mistakes have very real consequences, not all of them good. That is, Freud's (and others') illusion, which is part of the ontological rift, is not benign, like believing in Santa Claus or an imaginary friend. Freud's view is that nature should be feared (and perhaps resented, hated, etc.), which motivates human beings to control and, if possible, subjugate nature, which of course means other species.[40] But, if this is an illusion, what exactly do we fear and seek to control? It would be everything that falls under the heading of "nature," whether that entails the weather or other species (and othered human beings who are deemed to be closer to nature). Fear and anxiety foster conflict, separation, enmity, and so on, undermining cooperation and reciprocity. As for how destructive this illusion has been and is, consider the manifold examples of Western human beings' destructive subjugation and exploitation of other species (factory farms, industrial slaughterhouses, animal experimentation) and of the Earth (e.g., numerous dead zones in the ocean, mountaintop removal mining). It is not "nature" that is cruel, indifferent, and relentless but human beings of the rift who, out of fear and anxiety, seek to control and dominate (negative forms of power) other species and the environment.[41]Rosa (2020) describes the consequences of our attempts to control the world

> as alienation as opposed to adaptive transformation [Marx], as reification rather than revivification (Adorno and Lukács), as loss of world rather than gaining world (Arendt), as the world's becoming unreadable as opposed to comprehensible (Blumenberg), and as disenchantment as opposed to ensoulment (Weber).
>
> (p. 28)

Civilization

It is already apparent that Freud's view of nature is inextricably tied to the notion of civilization. In this section, I explore some of the attributes of Freud's view of civilization, such as its functions and sources. It is important to stress that Freud's view of civilization is, understandably, thoroughly Western in its origins and meanings, though Freud did recognize some of the downsides that he accepted as the cost of doing business in the context of the cruelties and indifference of nature. This said, the very Western notion of civilization from which Freud wrote is also the same civilization(s) that colonized and colonizes other peoples and lands (including the atmosphere), racialized and racializes othered human beings for the purposes of control and use, subjugated and subjugates women and excluded and excludes them from political processes, created apparatuses of nationalism and capitalism, and instrumentalizes other species and the land.

To begin, Freud (1927/1961a) remarked:

> Human civilization, by which I mean all those aspects in which human life has *raised* itself *above* its animal *status* and differs from the life of *beasts* … presents, as we know, two aspects to the observer. It includes on the one hand all knowledge and capacity that men have acquired in order to control the forces of nature and to extract its wealth for the satisfaction of human needs and, on the other hand, all regulations necessary in order to adjust the relations of men to one another.
>
> (pp. 5–6, emphasis mine)

There are several key features here. First, civilization has an inherent hierarchy when it comes to other species (I will say more below about "primitive" people). Second, as indicated above, civilization serves to control the forces of nature for the sake of protection as well as for the creation of wealth or, more broadly, for the well-being of civilized persons (see Freud, 1930/1961b, p. 92). Civilization's control over nature uplifts wealth, beauty, cleanliness, and order—features that "occupy a special position among the requirements of civilization" (p. 93).[42] Third, echoing Thomas Hobbes, Freud believed that civilization and, by implication, sovereignty are functionally necessary for regulating the social interactions of men, women, and children.

I mention Hobbes because his political philosophy appears to undergird Freud's view of civilization. Hobbes believed that for civilization or political society to function required citizens to submit to a leviathan for the sake of social order and obtaining some degree of freedom. Failing to do so would result in chaos, wherein life would be brutal and short (Ryan, 2012, pp. 430–435). In a similar way, as with Hobbes's citizen who has to sacrifice some measure of freedom to survive and flourish, Freud's civilized individuals sacrifice or renounce some of their desires in order to live in a society that provides stability and protection. While, for Freud (1927/1961a), nature "would not demand any restrictions of instinct from us, she would let us do what we liked" (p. 15), civilization does demand we collectively attenuate our instincts to obtain the desired ends of protection, cooperation, social belonging, wealth, beauty, and so on. For instance, civilization, according to Freud (1930/1961b), "restricts sexual life" by demanding an "indissoluble bond between one man and one woman" (pp. 104, 105). Of course, other social bonds are crucial to civilization, such as family, neighbors, and friends (pp. 108–109). This requires encouraging these relationships (e.g., love of neighbor, p. 109), as well as curbing instinctual aggression (pp. 111–116),[43] which, as history shows, has had mixed results.

But what does this view of civilization have to do with the ontological rift in Western philosophy? I have already noted the binary antagonistic juxtaposition between "nature" and "civilization." Now I want to turn to an unstated feature of Freud's view of civilization that imports the ontological rift into the very structure of civilized society. The indexes of Freud's works indicate that the notion of sovereignty is absent (Freud, 1999). The indexes contain no mention of sovereignty, which is curious because it lies at the heart of Western political philosophies and theologies. This absence, however, does not mean that sovereignty is indeed absent. I argue that it is central to Freud's view of civilization. Let me explain.

Freud makes use of the Greek story of Oedipus in developing his theory of psychosocial development. For instance, Freud (1918/1955b) accepted the view of "phylogenetically inherited schemata," of which the Oedipus complex is "the best-known member of the class" (p. 119). The story entails a king, Laius, and his newborn son, who is taken, on the king's order (apparently unbeknownst to Jocasta, his wife), to be killed so that Laius can secure his own rule into the future. Later, this son, now grown up, returns and unknowingly kills his father and marries his mother. Freud indicates that this family dynamic of killing (aggression) the father and marrying (libidinal instinct) the mother is a central, universal feature of human development. While there have been numerous critiques of this view, my point is that Freud unquestioningly imported patriarchal sovereignty into his anthropology and developmental theory, both of which he believed to be universal.

But it is not only a Greek political myth that Freud used to (unconsciously?) smuggle in the idea of sovereignty. Freud (1913/1955e), like Hobbes and Rousseau, created an imaginative story about the emergence of civilization (and religion). In *Totem and Taboo*, Freud told the story of the fictional father of a clan who had near total power and control over his sons, daughters, wives, and so on. The sons banded together and killed their father, thus becoming rulers themselves. Again, we note the unquestioned premise that (patriarchal) sovereignty is necessary for organizing society and later civilization—a society and civilization of command and obedience.

I wish to stress that Freud's notion of civilization (and his theory of development) unhesitatingly accepts a Western foundational political premise that sovereignty is necessary for organizing society. The notion of sovereignty vis-à-vis civilization requires a further, but brief, explanation of what the term means, its attributes, and its relation to the ontological rift. The notion of sovereignty has roots in ancient Greek philosophers such as Plato[44] and Aristotle (Grayling, 2019), though it was a French jurist and political philosopher, Jean Bodin, who, in the 16th century, set out to identify the key attributes of sovereignty instantiated in monarchies. Bodin (2009) identified four key features of sovereignty: supreme power (there is no superior), absoluteness, indivisibleness, and perpetuality. The king has no superior (except God) and is absolute in his rule;[45] his power and rule cannot be divided, and his rule is perpetual, handed down to his sons (in rare cases, daughters). This pre-Enlightenment formulation of sovereignty was not completely abandoned by Enlightenment political thinkers; it is notable in the political philosophies of Hobbes, John Locke, and, in the 20th century, jurist Carl Schmitt.

Carl Schmitt (2005), in his study of law-making and law-preserving actions of the state, reinterpreted Bodin's work. He boldly stated that the central feature of all forms of sovereignty is the state of exception. "The sovereign," he wrote, "is he who decides on the state of exception" (p. 5). Besides the fact that Schmitt carried forward the idea of sovereignty embodied in an ostensibly male person, there are two key points here. The sovereign is the one who makes the decision, which might awaken our memory of a democratically elected president who said he was the decider in chief or the Supreme Court's 2024 decision regarding presidential immunity. Schmitt went on to claim that a

> sovereign produces and guarantees the situation in its totality. He has a monopoly over this last decision. Therein resides the essence of the state's sovereignty, which may be juristically defined correctly, not as a monopoly to coerce or to rule, but as a monopoly to decide.
>
> (p. 13)

The second and more complicated feature here is the state of exception and its relation to the law. The phrase "state of exception" means that the sovereign, as we see in the quote above, can, because of absolute power, decide to set aside laws according to his (historically, sovereigns are mostly male) own assessment of the requirements of a given situation—*extremus necessitatis casus* (p. 10). Bodin's notion of absolute power and authority, then, is seen in Schmitt's notion of the state of exception.

The relation between the law and sovereignty's state of exception needs further elaboration. For both Agamben and Schmitt, the "sovereign exception is … the condition for the possibility of the juridical order, for it is through the state of exception that sovereignty creates and guarantees the order the law needs for its own validity" (DeCaroli, 2007, p. 53). At the same time, "The state of exception exercises the law as force" (Colebrook & Maxwell, 2016, p. 54). Agamben, then, takes on Schmitt's idea that the sovereign stands outside of, but belongs to, the juridical order, which is a paradox. The paradox is yoked to questions of the relations between sovereignty, the law, and violence. "The paradox of sovereignty," Agamben (1998) writes, "consists in the fact that the sovereign is, at the same time, outside and inside the juridical order" (p. 15). As Agamben (2005) writes elsewhere, "The state of exception appears as the legal form of what cannot have legal form … if the law employs the state of exception—that is the suspension of the law itself" (p. 1). Put differently, "As a figure of necessity, the state of exception therefore appears as an 'illegal' but perfectly 'juridical and constitutional' measure" (p. 28). Sovereignty, then, "is neither external nor internal to the juridical order" (p. 23). While paradoxical,

> The sovereign exception is, for both Schmitt and Agamben, the condition for the possibility of juridical order, for it is through the state of exception that the sovereign creates and guarantees the order the law needs for its own validity.
>
> (DeCaroli, 2007, p. 54)

This stated, the paradoxical[46] nature of sovereignty is evident in that the "work of sovereignty precedes the law, creating a regular 'frame of life,' which the law preserves and codifies but does not instantiate" (p. 50). Recall Freud's early tribal father and his rule over other men and women. As sovereign, his rule precedes the law, but it is later codified in patriarchal apparatuses that preserve the sovereign, who, in turn, can suspend laws.

There are several related features of sovereignty's state of exception that reveal the presence of an ontological rift. First, sovereignty is universalized or ontologized in Western thought. Aristotle and Plato constructed political philosophies that posited the idea of sovereignty as existentially necessary for the polis, for civilization. In the Abrahamic traditions, God is sovereign, which later legitimized male sovereignty or what Murray Bookchin (2005) calls "epistemologies of rule" (p. 152). Second and relatedly, sovereignty as an ordering premise of civilization depends on political coercion and violence. John Lechte and Saul Newman (2015) contend that the "law always articulates itself through violence which both preserves its boundaries and exceeds them, and violence always establishes a new law. … [V]iolence is at the very origins and foundations of the law" (p. 128). In brief, the "Law," they point out, "is never free from violence" (p. 173). This echoes Foucault's view that the "the law is born of real battles, victories, massacres, and conquests, which can be dated and which have their horrific heroes; the law was born in burning towns and ravaged fields" (as quoted in Oksala, 2012, p. 40).

There is, however, a deeper connection between law and violence. In DeCaroli's (2007) reading of Agamben, the "law is not the essential function of sovereignty" (p. 50). "The work of sovereignty," he writes, "precedes the law, creating a regular 'frame of life,' which the law preserves and codifies but does not instantiate" (p. 50). What precedes the law is violence or the threat of political violence. Freud's fictional story about the origins of civilization bears this out. The killing of the tribal father is an act of political violence aimed at overthrowing the sovereign. Yet, this political violence is necessarily preceded by the violence and the threat of violence the father must have employed to subjugate his sons, daughters, and other members of the tribe. Political violence establishes the sovereign, and then the sovereign and sovereign classes create laws (and stories) to legitimate the ruler and ruling bodies, all of which is never far from the threat of political violence.[47]

Third, sovereignty and political violence (or the threat of political violence) are fundamentally exclusionary and depersonalizing. Laius exercises his sovereignty by having his son killed, which is an ultimate form of the state of exception, political exclusion, and depersonalization. Agamben's (1998) notion of bare life, Fanon's (1952/2008) idea of a zone of nonbeing, and Orlando Patterson's (1982) social death are terms that point to the results of political violence and depersonalization toward citizens and noncitizens. Exclusion can include the restriction or denial of political participation (e.g., exclusion of women,[48] enslaved persons, and persons of color; imprisonment), expelling persons constructed as noncitizens (e.g., deportation), and legal forms of killing (e.g., police killings, executions).

I mentioned above another form of sovereignty's exclusion. Recall that Agamben (2004), Latour (1993), and Derrida (2008) note that there are, in the West, "entirely distinct ontological zones: that of human beings on the one hand; that of nonhuman on the other" (Latour, 1993, pp. 10–11). Other-than-human species, since they are deemed to lack political agency and personhood, are seen as not belonging to the political realm except in terms of their use value (through the depersonalization of factory farms, industrial slaughter factories, and use of animals for war). Nevertheless, sovereignty applies in that all human beings are believed to rule over other species. Human beings, in Western political and theological traditions, are constructed as superior, legitimating their dominion over and violence toward other species and the Earth. To return to Freud (1927/1961a), his notion of civilization rarely mentions other species, though when he does it is in the contexts in which "[civilized] human life has raised itself above its animal status and differs from the life of beasts" (p. 6).[49] Of other species, for instance, "Nobody talks about the purpose of the life of animals, unless, perhaps, it may be supposed to lie in being of service to man" (Freud, 1930/1961b, p. 75), which demonstrates the use value of other species. Let me stress that varied forms of exclusion, which are evidence of the rift, are based in innumerable examples of political violence and coercion; these are legitimated in philosophical and scientific apparatuses that are premised on the idea of sovereignty as necessary for civilization (and human relations to other species and the Earth).

In summary, Freud's view of civilization is inextricably linked to his understanding of nature and the dangers nature poses to human well-being. At some point in human evolution, in Freud's schema, we gathered in groups and developed apparatuses of culture and civilization that secured our need for protection against the forces of nature. This view of civilization, I argued, imported the fundamental and often unquestioned premise that sovereignty and its disciplinary regimes are necessary for the stability and security of civilized society. This importation included the production of varied forms of the ontological rift, which is evident in the exercise of political coercion and violence aimed at excluding some persons or groups (e.g., women, people of color) and, of course, other species. I argued, in other words, that Freud's view of civilization, which is inextricably joined to Western philosophical premises regarding sovereignty, is similarly founded on political violence, depersonalization, and exclusion. Like his view of nature, civilization represents the presence of the ontological rift, which is manifested in the beliefs in superiority and inferiority, attendant depersonalizing epistemologies, the construction of apparatuses of sovereignty, and the exercise of negative power over other human beings or other species.

Indigenous Peoples and the Ontological Rift

Western civilization is a semiotic fortress that has excluded and exploited other species and othered human beings for millennia. "Civilized" sovereign nations such as Britain, France, Germany, and the United States have constructed social imaginaries that justify and legitimate racism, imperialism, and the colonization of most of the Earth and its peoples (Mills, 1997, 2017; Robinson, 1983; Woods, 2017). These exploitative social imaginaries of civilization and sovereignty made their way into the various human and physical sciences (Go, 2016), which, as noted above, were (and, in some cases, are) implicated in producing and reproducing the ontological rift. This is certainly the case for early anthropologists who sought to study other peoples—anthropologists Freud relied on while developing his thoughts about human development and belonging. Freud understandably made use of anthropological studies of his day; this is not a criticism. Rather, I am highlighting that the ontological rift was also evident between human beings, namely, between civilized and so-called primitive or uncivilized peoples. I have mentioned this above, but here I briefly explicate Freud's use of these appellations because they reveal the presence of the rift in his theory of psychosocial development and his view of human evolution. This is important for the next chapter's reimagining of psychosocial development and epistemology without the rift.

Freud's earliest works to his latest make numerous references to primitive peoples and races.[50] The term "primitive" is not simply associated with putatively uncivilized peoples (see Brickman, 2018; Swartz, 2018). Freud uses the term when referring to psychosocial development and concepts such as id, primary process, irrationality, instincts, and religion. Freud's view of psychosocial development is closely linked to his beliefs regarding human evolution, which makes sense because, as Frank Sulloway (1992) notes, in the late 19th century it was commonly believed among evolutionary theorists that ontogeny recapitulates phylogeny. This means that "the development from fetus to adulthood (ontogeny) provides a brief recapitulation of the entire history of the race (phylogeny)" (p. 259). For instance, Freud (1913/1955e) argued that there are three stages in human evolution, namely, animism, religion, and science (p. 77). He believed "primitive" peoples and children operate in the animistic phase—a "lower" stage of development and evolution (children are not yet civilized). While I will say much more about animism in the next chapter, for now let me say that Freud (1927/1961a) linked animism to narcissism in psychosocial development (p. 90), omnipotent thinking (p. 85), and the personification of the forces of nature and other species (p. 22).[51] Freud considered "primitive peoples," who have not yet achieved the "higher" stage of scientific thinking, to be trapped in the animistic stage. He wrote,

> Primitive man has no choice, he has no other way of thinking. It is natural to him … to project his existence outwards onto the world and to regard every event which he observes as a manifestation of beings who at bottom are like himself. It is his only method of comprehension.
>
> (1930/1961b, p. 122)

Of course, children living in a civilized nation are only trapped until they move to the next stages, namely, religion and, hopefully, even "higher" to science. The "neurotic" (Freud, 1913/1955e, pp. 86–89) and uneducated people (Freud, 1933/1964c, p. 122), like primitive peoples, are, for Freud, also captive to this earlier stage of omnipotent thinking.

It is no surprise to anyone who has read even a little of Freud that he privileged science as an epistemological method above all others. To be sure, Freud acknowledged that all human beings have a primitive core (even scientists) or go through an animistic stage of development; this may seem to bridge the rift between civilized human beings and "savages," but in reality it does not. The implicit taxonomic hierarchy embedded in his formulations is one associated with Western philosophical thought that privileges reason above what is deemed to be irrational or primitive. This hierarchy of higher and lower (superior and inferior) (1) leads to a split between personal and instrumental epistemologies and (2) is integral to the social imaginaries of imperialistic nation-states that constructed and construct Indigenous peoples as primitive or savage—as lacking reason and, therefore, in need of being civilized. Ethnic cleansing, the destruction of Indigenous cultures, enslavement, and the exploitation of Indigenous peoples are indisputable signs, symptoms, and effects of the "logic" of the apparatuses that produce the ontological rift.

I add that Freud (1913/1955e) himself acknowledged a consequence of the rift when he mentioned "our attitude towards the mental life of children, which we adults no longer understand and whose fullness and delicacy of feeling we have in consequence so greatly underestimated" (p. 99). Recall my mention of Frank Linderman, who spent decades with the Crow people and who acknowledged that, in the end, he could not quite understand them. The connection between Linderman and Freud is their reliance on Western epistemologies that are rooted in the ontological rift. The use of Western social imaginaries to understand Indigenous people and children led to misunderstandings, at best, and traumatic relations at worse (e.g., Bunge, 2001; Miller, 2002). Freud, like Linderman, sought to understand human beings, but their comprehension was occluded when it came to children and Indigenous people, despite their sympathies and desire to understand.

A bit more needs to be said about this rift in relation to Indigenous peoples. For Freud, Cartesian–Baconian science was the epistemological method for investigation, and it objectified the targets of investigation. This meant that, in Freud's language, animistic epistemologies were of little or no use in furthering knowledge since they were captive to omnipotent thinking. To be sure, Freud the scientist would have acknowledged that he, too, had a primitive core or side, but he believed he overcame this by way of adopting scientific thinking and methodologies. This means that Freud split off animistic and religious epistemologies from scientific methods. In privileging science and splitting it off from animistic thought, Freud was unable to see and understand animistic epistemology's evolutionary and ecological importance or pragmatic utility for learning about the environment and its inhabitants for the sake of survival and flourishing. In short, Cartesian–Baconian science serves as an apparatus of the ontological rift, which means it is not capable of acknowledging the ecological value of so-called primitive epistemologies. Moreover, his theory is incapable of integrating these epistemological approaches, which I do in the next chapter. There is, finally, in the midst of privileging science, a subtle disdain for "primitive" peoples and adults who continue to operate out of animistic thought. Privileging science and the corresponding disdain are, in my view, symptoms of the ontological rift.

The term "symptom" brings me to some concluding thoughts concerning the rift and Freud's view of nature, civilization, and "primitive" people vis-à-vis the ontological rift. The architectural flaw in Western philosophy is, in one sense, a symptom. For Freud (1926/1959b), a symptom involves displacement, wherein the symptom resembles but is different from what it symbolizes (pp. 110–111). The aims of symptom formation are gratification and repression (pp. 94–95). If the architectural flaw is evident in Freud's work and his view of "primitive" peoples, what, then, is repressed and gratified? Briefly, a possible answer is that "primitive" people and other species are like us but unlike us. What is repressed in Cartesian–Baconian science is our capacity to personalize other species—to recognize them in terms of their singularities, which is not the same as anthropomorphizing them. What is gratified is the need or desire to know these other species. However, this knowing is objectifying and instrumentalizing, which *separates* the knower from the object. This view leaves open the question, why repress our capacity to personalize or, more broadly, to animate?

If we return to Freud's views of nature and civilization, we can discover a partial answer to this question. Freud continually used the term "control" when discussing nature and civilization. The history of the Western sciences is one of controlling nature (within limits, of course), but this is in the service of protection, wealth, and so on. What is displaced or repressed is our collective helplessness in the face of the forces of nature, and what is gratified is our sense of agency, power, and control. But what if we were to suggest that "helplessness" in relation to the forces of nature functions, in Freud's work, to screen or repress experiences of existential insignificance? I think there is evidence for this interpretation. First, the hierarchy related to psychosocial development and evolution provides us with a sense of gratification that we are not children, uncivilized, and so on. This gratification is linked to senses of self-esteem and self-confidence that rely on the beliefs in the superiority of civilized persons and the inferiority of Indigenous persons—like but unlike. Yes, Freud said we all have a primitive aspect to us, but he believed this is to be overcome through psychosocial development or evolution or science. Beliefs in superiority and inferiority are understood further in terms of significance and insignificance. If existential insignificance is repressed, insignificance is projected onto Indigenous people and other species, though they may retain use value.

Second, Freud's considerable writings represent, in part, a desire for public significance. For Freud, death was a reality he thought of often, and "death" is the ultimate signifier of existential insignificance. In the face of death, most human beings construct social imaginaries and practices (e.g., memorials) that will outlast them—that will embody their significance after their death. Freud created psychoanalysis and wrote voluminously right up to his death, but in the end, when humans become extinct, psychoanalysis and Freud will slide into the abyss of existential insignificance. Put another way, I suggest that, while he recognized the experience of helplessness in the face of death, Freud's theories can be seen as his attempt to displace the painful reality of existential insignificance onto other species and Indigenous people.

I wish to stress that this does not simply pertain to Freud. Western philosophies and theologies have replicated the ontological rift. The notion of "symptom" and the repression of existential insignificance apply to these imaginaries. Western taxonomic hierarchies, beliefs in the superiority of Western civilized persons, the objectification and instrumentalization of nature and othered peoples, and social-religious and philosophical imaginaries that ontologize and eternalize a group of human beings are symptoms of Western subjects' anxiety regarding the reality of existential insignificance that is repressed. Gratification is obtained by holding on to their significance at the expense of the insignificance of other species and othered human beings.

Finally, a symptom is a semiotic construction, which can be understood as a narrative spelling out (Stern's weak dissociation) that accompanies unconscious features. Freud's innovation was to get the symptom to speak, which means that what was previously unconscious is now conscious. Freud's construction of civilization, nature, and Indigenous peoples can be understood as a symptom—as semiotic constructions that result in a normative, historical, developmental, and political unconscious. Civilized human beings are rendered normative, while other species and Indigenous (nonnormative) people are seen to be lower on the taxonomic hierarchy of significance. Freud's imaginative political history of the rise of civilization, for instance, makes no mention of other species—a consequence of the ontological rift. His developmental perspective, as Harold Searles (1960) notes, does not include the importance or necessity of other-than-human beings or what Searles calls the nonhuman environment. Putting Freud's notions of nature, civilization, and Indigenous people on the couch creates the possibility for the symptom to speak of the architectural flaw of the ontological rift, which is no longer hidden.

Conclusion

Like any great innovator, Freud made use of the intellectual tools at hand to create theories and ideas that would influence generations of analysts, literary theorists, philosophers, and so on. The tools Freud relied on were grounded in Western philosophies and Cartesian–Baconian sciences that contained an architectural flaw, namely, the ontological rift between human beings and other species (and othered human beings). As noted above, Freud was aware of this rift but did not carry this insight into his analysis of nature, civilization, and Indigenous peoples. Indeed, in this chapter, I have argued that Freud's views of nature, civilization, and Indigenous peoples illustrate the presence of the ontological rift throughout his work. I have also considered Freud's views and the Western ontological rift as symptoms that repress existential insignificance while at the same time gratifying our desire for self-esteem, agency, and control. The next chapter employs these insights to construct a developmental theory that overcomes the rift.

Notes

1 Philosopher Lewis Gordon (2023) notes that the word "science," "although meaning knowledge, reveals much in its etymology. It is a transformation of the Latin infinitive *scire* ('to divide'—think today of 'schism'), which, like many Latin words, also shares origins with ancient Greek words, which, in this case would be *skhizein* (to split, to cleave). Oddly enough, this exercise in etymology is indication of a dimension of epistemological colonization" (p. 200). This is especially evident given the rift between human beings and other species.

2 It is important to mention that, in Western history, philosophers and others have advocated including other species in discussions about ethics and justice. Stephen Newmeyer (2005) addresses the work of Plutarch and his argument about the rights of animals. A couple of centuries after Plutarch, Neoplatonic scholar Porphyry (1823), writing in the third century, argued it was our moral duty to extend justice to other animals (Book 1.11–44). Centuries later, Jeremy Bentham, Johannes Kniess (2018) writes, argued for the welfare of animals. Also, Stephen Puryear (2017) explores the 19th-century work of Arthur Schopenhauer and his concern for the rights of animals. More recently, philosophers such as Peter Singer (1975, 2023), Andrew Linzey (2009), and Martha Nussbaum (2022) have explored thorny ethical issues regarding human beings' treatment of other species. Add to this the Romantic literature that posits that "the modern development of rationality and instrumental reason has alienated us both from our own emotional nature and from Nature we live in" (Taylor, 2024, p. 96). The Romantic poets, Charles Taylor argues, believed it is "our destiny to heal both of these rifts together" (p. 96). In the area of science, in the 19th century, Alexander von Humboldt and others made "no concessions to Cartesian-Baconian objectification of nature" (p. 261). See also Wulf (2015).

3 The concept of "nature" has a great deal of plasticity in routine or academic discourses. It can refer to all living and non-living things. It can refer to an individual species of animal (e.g., human nature, nature of horses, cows). Aristotle used the term "second nature" to refer to the state of perfection (Khurana, 2023, p. 36). Millennia later, Hegel, according to Axel Honneth (2023), referred to human beings' second nature when depicting ethical customs, civil society, and human freedom (p. 19). Nature can also mean something decidedly different from human beings, as in Mother Nature, though one can acknowledge at the same time that we are part of Nature. Lynne Segal (2023) rightly notes that "nature" "is implicitly anthropocentric" (p. 159). This discourse reveals not simply the semiotic capacity to differentiate between this and that object(s) or to differentiate between human beings and the larger world. It also reveals that we enact distinctions as if they are real instead of mere constructions or abstractions that help or hinder our way of navigating and dwelling in the world.

4 A recent book by Richard Boothby (2023) on Lacan and religion is another example of how psychoanalytic theories unwittingly import the ontological rift. Besides the repeated use of the term "archaic" in referring to earlier human beings and early development, Boothby argues that Lacan's notion of the signifier provides "a measure of escape, of achieving a margin of independence, from the gravitational bond of the other" (p. 46). This follows on the heels of Boothby's mention of Hannah Arendt's reference to the achievement of *Sputnik*. She uses the words of an American reporter who said this achievement was a "step toward escape from men's imprisonment to the earth" (p. 45). The reporter quotes a Russian rocket scientist who said, "Mankind will not remain bound to the earth forever" (p. 46). Then Boothby references a primatologist who remarked, "When I am with the bonobos, I feel like I have something that I shared with them long ago

but forgot. As we've clothed ourselves, we've gained a wonderful society [civilization], but we lost a kind of soul-to-soul connection that they maintain" (p. 46). To view the Earth as a prison that human beings seek to escape is a Western preoccupation, one that is tragic since the Earth is our material home that we are destroying. And to "forget" our connection to other species and the Earth signifies the presence of the ontological rift. I would also mention that Boothby later comments that "a child is always a stranger in a strange land" (p. 46). It is not clear that this is universally the case, though homelessness is a perennial theme in Western philosophies and theologies. Having read Arendt's work and returned to her prologue, Boothby misreads her in his aim to connect Lacan to Arendt. Arendt (1958) remarks that Western human beings have not only rejected "a god who was father of men in heaven" but also "an even more fateful repudiation of an Earth who was the Mother of all living creatures" (p. 2). In my interpretation, Arendt is pointing to a symptom of the ontological rift, namely, the rejection of our dependency on the Earth for life.

5 Some of these same criticisms could easily be levelled at Freud. Freud proposed the ideas of the Oedipal complex and the death instinct, which provided universal, essentialist explanations. Moreover, as I explain below, there is a tinge of animistic thinking in Freud's work, which is not necessarily a criticism, but from his point of view it would be.

6 Psychoanalyst Donna Orange (2009, 2017) has been an advocate for engaging philosophy in psychoanalytic discourse and theorizing—a view I obviously agree with.

7 Freud also relied on anthropologists of his day (e.g., Batchelor, Boas, Frazer, Howitt), though usually as support for his theories and not as critical corrections of his scientific or philosophical views. These anthropologists, in other words, carried Western negative biases regarding so-called primitive peoples, which fit neatly with Freud's developmental theories, though with some qualifications. The decolonizing anthropologists of the late 20th and early 21st centuries are critically aware of these biases embedded in Western philosophies and sciences. I use them as critics and correctives of Western philosophies and sciences.

8 By "Western persons," I mean persons who have internalized the narratives and practices of European anthropologies. There are people and groups in the West, such as Indigenous peoples, whose anthropologies do not produce the ontological rift.

9 Roland Faber (2023), exploring the work of Alfred North Whitehead, locates the philosophical source of the rift in Plato's philosophy. He writes that "the platonic divide becomes unmasked as an abstraction from the process of experience and not its condition" (p. 18). Whitehead also said that all Western philosophy is but a footnote to Plato's work, which suggests that this abstraction, this platonic divide, is evident in Western philosophies and their attendant social imaginaries such as the human sciences.

10 Derrida (2008) does not lay the blame for this abyssal rupture on the Cartesian mechanistic view of animals or on Kant's "hatred and revilement" of other-than-human species. Rather, he rightly accuses "the Judeo-Christiano-Islamic tradition of a war against the animal, of a sacrificial war that is as old as Genesis" (p. 101).

11 Scholar Murray Bookchin (2005) argues that the domination of other human beings emerged long before human beings began to think of dominating nature. Considering Freud's imaginative recreation of the origins of society in *Totem and Taboo* (1913/1955e), wherein the father of the clan dominates others only to be killed and overthrown by his progeny, he may have agreed with Bookchin. Regardless, I am inclined to agree with Bookchin and add that the social imaginaries that legitimated the domination of one group of human beings over others served to extend this way of dwelling in relation to what we in the West call "nature."

12 I will say more about this insight, but for now I am arguing that Freud did not see the epistemological evolutionary value of constructing other animals as persons—unique, valued, inviolable, responsive subjects. Freud recognized this human arrogance and the resulting rift, but, because he retained (1) a Cartesian–Baconian view of science and its objectifying and instrumental epistemologies and (2) a belief in the necessity of civilization, which is founded on hierarchal relations of command and obedience (see Bookchin, 2023), for human survival, he was not able to make use of this insight to amend his anthropological theories.

13 Francis Bacon (1561–1626) believed that "the practical aim of improving humanity's lot [depended on] the increased understanding and *control* of nature" (Grayling, 2019, p. 197, emphasis in original). Similarly, citing Carolyn Merchant's work, Jean-Baptiste Fressoz (2015) stated that, in the Renaissance and up to the scientific revolution, "our planet was conceived as a living body. … The Earth was a mother that had to be respected. The scientific revolution and the emergence of capitalism led to an inexorable decline of organic cosmology. Nature became a vast mechanism to be explained, mastered, and exploited" (p. 76). Mastery or control of nature, which I will say more about below, relies on instrumental, objectifying epistemologies that function to produce and maintain the ontological rift, ostensibly for the benefit of human beings alone.

14 Aristotle (1971), and later Hobbes, Bodin, and Schmidt, believed that sovereignty or ruling was a necessary feature of life. "Ruling and being ruled," Aristotle wrote, "not only belongs to the category of things necessary, but also to things expedient. … This characteristic is present in animate beings by virtue of the whole constitution of nature, inanimate as well as animate" (p. 12).

15 For Macmurray (1961), personal recognition is both a matter of intention and a matter of fact. This paradox demonstrates Macmurray's desire to hold on to the philosophical perspectives that a human being qua person is an existential fact and that human beings can intentionally, by way of freedom, deny this fact, which he noted in varied forms of depersonalization evident in human affairs. In addition, MacMurray argued that object knowing (which includes technological, instrumental epistemologies) appears first in development and is the basis of all knowing, including personal knowing. That said, to recognize the individual as a person requires placing object knowing as secondary. The only time object knowing is ethically primary is when it is for a short time and clearly for the sake of the other person. A psychologist or physician diagnosing a patient uses forms of object knowing, and this is ethical if it is for the sake of benefiting the patient. One final comment: to acknowledge the singularity of another being is similarly a matter of fact and a matter of intention. The singularity of another being is an existential matter of fact in the sense that an individual being is unique, values itself, and possesses some level of integrity. Singularity as a matter of intention entails human agency, perception, and treatment of the being as singular. Of course, there are innumerable examples in our everyday lives where we lack the matter of intention, resulting in varied kinds of exploitation and destruction of other beings. Finally, the experience of singularity, which for human beings initially depends on the personal recognition or knowing of good-enough parents, is understood as embodied senses of Sartre's notions of being-in-itself and being-for-itself.

16 Lewis Gordon (2023) notes that, in Aristotle's world, the "male stands as the formal constitution of human reality; [the] female lays bereft as an undeveloped male and, consequently, an undeveloped human being, a less-than-human being" (p. 100). Men, then, are considered "normative," which, as noted below, means women are nonnormative and largely written out of politics and history. I add here that racism establishes the white male and female as normative and persons of color as less-than-human.

17 Macmurray (1961) limits his view of personalization to human beings, which means the idea that other species could be constructed as persons lies outside his philosophical perspective. This omission is, in my view, the result of the presence of the ontological rift in his neo-Kantian philosophy.

18 While the philosophical notion of freedom is complicated, let me note that in relation to the ontological rift scholars recognize that freedom is actually connected to the rift. Erich Fromm viewed this as negative freedom because it represents separation and alienation with regard to nature. Positive freedom aims to overcome this alienation to form a union with nature (see Thorpe, 2020, p. 170). Reading Marx, David Harvey (2016) also remarks that estrangement from nature gave rise to "emancipatory forms of knowledge (such as science); but it also poses the problem of how to return to that which consciousness alienates us from" (p. 205). In my view, the ontological rift is part of Western thought. We may find that other cultures do not rely on alienation from nature vis-à-vis consciousness, agency, and freedom. In other words, Marx, Fromm, and Harvey seem to think that estrangement from nature is universally necessary for consciousness when it is simply a factor of Western philosophies and theologies.

19 Political theorist Carl Schmitt recognized that "all significant concepts of the modern theory of the state are secularized theological concepts" (as cited in Brown, 2010, p. 59). This is analogous to Lacanian psychoanalysis's focus on "lack," as Jaques Derrida (2008, p. 122) points out in his critique of Lacan's Western, secularized Christian anthropology. Derrida writes, "It is not just a matter of asking whether one has the right to refuse the animal such and such power (speech, reason, experience of death, mourning, culture, institutions, technics, clothing, lying, pretense of pretense, covering of tracks, gift, laughter, crying, respect, etc.—the list is necessarily without limit, and the most powerful philosophical tradition in which we live has refused the animal all of that). It also means asking whether what calls itself human has the right rigorously to attribute to man, which means to attribute to himself, what he refuses the animal, and whether he can ever possess the pure, rigorous, indivisible concept, as such, of that attribution" (p. 135).

20 The idea of personhood within the context of human beings is complicated and often used as a political cudgel in denying rights to women, as in the case of the abolition of medical care. I am not going to wade into this debate but rather will focus on the philosophical and psychological use of this term in relation to other species. I add here that the recognition by Indigenous peoples of the personhood of animals, rivers, etc., does not mean that they do not use other animals or kill them. Nevertheless, the idea of another animal being a person (such as a bear-person) alters one's dispositions and perceptions.

21 Agamben (1999), relying on but amending Aristotle, views all living beings in terms of actualizing their potentialities. I will say more about this in the next chapter, but for now it is important to mention that the movement from potentiality to actualizing this depends, for human beings, on a personal recognition that founds caring actions.

22 The racial contract, according to Charles Mills (1997), comprises a "political system, a particular power structure of formal and informal rule, socioeconomic privilege, and norms for the differential distribution of material wealth and opportunities, benefits and burdens, rights and duties" (p. 3). Mills points out that this contract is political, moral, and epistemological (p. 9) and that it "is clearly historically locatable in a series of events marking the creation of the modern world by European colonialism and the voyages of 'discovery' now increasingly called expeditions of conquest" (p. 20). Mills adds, "We live in a world which has

been foundationally shaped for the past five hundred years by realities of European domination and the gradual consolidation of global white supremacy" (p. 20).

23 Winnicott (1971) challenged the tendency to consider illusions (mistaken or misleading beliefs) as negative. He believed illusions could be benign and even developmentally necessary (e.g., illusions associated with transitional objects). One could argue that the apparatuses of the ontological carry a number of illusions. However, I opt for the stronger term *delusions* because the beliefs of the rift are existentially false. Human beings are not superior, nor are they disconnected from or beyond nature. There is nothing in existence that confirms human superiority. Indeed, human extinction, which will eventually occur, disproves human superiority, as well as beliefs in human dominion.

24 Carlo Strenger (2011) repeatedly refers to human fears of an indifferent universe and human beings as tragic creatures. Yet, I contend that the "universe" is a sign that represents the whole of creation. The universe is not an entity and cannot be indifferent. Only human beings and other semiotic beings can be indifferent, in the sense of not recognizing the significance of this or that object. I would also like to push back on the universal idea that human beings are tragic creatures. I certainly agree that persons of the ontological rift are tragic creatures because their way of dwelling in the world is leading to human extinction. I am not sure that Indigenous peoples, who do not produce apparatuses of the rift, view themselves as tragic, though they certainly have suffered tragedies at the hands of Western colonizers—colonizers who produce(d) and maintain(ed) the rift.

25 Ernst Becker (1997) argues that much of Western society is geared toward the denial of death. Freud would have agreed. "Death" is a sign that denotes the end of all signifying operations. We fear death, as Massimo points out, because it represents existential insignificance. When we deny death, we are also denying our existential insignificance.

26 Winnicott posited that infants initially possess the illusion of impotence, which gradually undergoes disillusionment by way of parents' moving away from being preoccupied with meeting infants' assertions. The belief in and experience of omnipotence, however, are transferred to the transitional object, which in adulthood can be seen, for Winnicott, in religion, science, and other cultural activities. An entire article or perhaps a book could address the varied forms of human beliefs in omnipotence in human life; this topic, therefore, cannot be taken up here. That said, we can approach the idea of human beliefs in omnipotence by way of their impact on how human beings dwell with each other, with other species, and with the environment. The ontological rift represents, in my view, beliefs in omnipotence that have to do with *power over* othered human beings, other species, and the Earth. This is evident in the Abrahamic religions and political philosophies, as well as in the Cartesian–Baconian sciences. Many Indigenous peoples possess omnipotent beliefs characterized by shared precarity, mutuality, reciprocity, and cooperation (see McDonough, 2024; Trosper, 2022).

27 I prefer depersonalization because personalization represents a kind of knowing that extends to other species, as is evident in many Indigenous societies (Viveiros de Castro, 2017). Dehumanization refers solely to other human beings. We would not say a bear is dehumanized, because it is a bear. But we could say it is depersonalized in that the bear's inherent value, uniqueness, inviolability, and agency are denied.

28 This is not the case in many Indigenous narratives where other species are part of the story and may be the focus of the story. See Kerven (2018), Deloria (1997, 2003), and Erdoes and Ortiz (1984).

29 Gordon (2023) makes a similar claim, though he focuses on apparatuses that marginalize and oppress othered human beings (p. 158).

30 This is not the case universally. Bryan Rousseau (2016) reports that the New Zealand government has political representatives or officials who hold office to represent the rights of the land and all its inhabitants. There are others who argue that other species should be included in political thinking (e.g., Meijer, 2019).

31 The Psychoanalytic Electronic Publishing library comprises over 151,000 articles and 138 books. The phrase "other species" brings up 500 (.0033%) hits, and "other animals" brings up 782 (.0052%) hits. "Human beings" has 11,500 search results (.078%), "men" nearly 28,000 (.19%), and "women" 25,654 (.173%). This is not definitive proof of the relative absence of other species in theorizing about human psychosocial development, but it is instructive.

32 In the Psychoanalytic Electronic Publishing library, rats are mentioned nearly 1,500 times, cats 1,000, dogs 2,225, and horses 1,528. It would be interesting, though time-consuming, to explore the contexts in which these animals are mentioned.

33 Sartre's notion of bad faith parallels Henrik Ibsen's (2009) play *The Wild Duck*, wherein Relling uses the term "life-lie." That is, we adopt varied constructions of the world that are, in the end, not simply fictions but lies. The ontological rift, in my view, is a life-lie.

34 Richard Twine (2024, p. 6) cites research indicating that over 73.1 billion farm animals were slaughtered in 2020. Factory farms and industrial slaughterhouses are obvious signifiers of the rampant human apparatuses that assign insignificance (or mere use value) to the animals that are excluded from our ethical considerations.

35 See footnote 15 in this chapter for some of the problems with constructing the universe as indifferent.

36 Implicit here is the notion that communal life depends on civilization, which by implication means that sovereignty is necessary for social life. This Hobbesian view implies that those who are deemed to be uncivilized are subject to the whims and cruelties of nature and cannot establish communal life. This is an illustration of the rift and the logic of colonization.

37 Freud (1927/1961a) also used the phrase "humanization of nature" (p. 22). This implies anthropomorphizing nature, which is decidedly different than personal knowing. Personal knowing, which I will say more about in the next chapter, simply means recognizing the object as a unique, valued, inviolable, responsive subject.

38 This view is not necessarily the case among scientists who do recognize and respect the singularities of other species (e.g., Dian Fossey, Jane Goodall, Danna Staaf).

39 Rosa (2019), in addressing climate change, notes that many of us continue to anthropomorphize nature. "I wish only to demonstrate here," he writes, "the persistent interpretations of events as *nature's revenge, nature striking back*, or *the scream of abused nature*" (p. 275, emphasis in original).

40 Freud (1930/1961b) wrote of "this *subjugation of the forces of nature*, which is the fulfillment of a longing that goes back thousands of years." He added that this control and subjugation "has not [led to an] increase [in] the amount of pleasurable satisfaction which they may expect from life and has not made them feel happier" (p. 88, emphasis mine). This observation, which I suspect is accurate for Western human beings who live lives of quiet desperation, is worth exploring. Could it be that the ontological rift and the instrumentalization of "nature" create experiences of alienation from "nature"? Charles Taylor's (2024) in-depth examination of the Romantic poets and their desire to heal the rift lends credence to this. See footnote 3 in this chapter.

41 In his discussion of the Frankfurt School, Malte Ibsen (2023) notes that Theodor Adorno wanted us to "practically reorient our relationship to nature in a mode

beyond instrumentalization" (p. 134). Western instrumentalization of nature, Adorno noted, has caused cataclysmic damage, and this happened before people recognized the harms associated with climate change. I add here that "nature" is again a construct that refers to human beings' relation to the environment.

42 Control over nature is repeated in a number of Freud's works, which indicates its centrality in his anthropology—an anthropology that reproduces the ontological rift.

43 Freud viewed human beings as possessing an aggressive instinct that is partially tamed by civilization. He tended to view "primitive people" as more aggressive and crueler than civilized people (who have the potential for devolving into primitivity). He wrote, "Primitive man ... was no doubt a passionate creature and more cruel and more malignant than other animals. ... Hence the primaeval history of mankind is filled with murder" (1915/1957b, p. 292). I believe civilization corners the market on aggression, cruelty, and violence; a cursory reading of history will attest to this.

44 Karl Popper (2002) argued that Plato's fundamental philosophical question regarding the polis was who should rule. This, he contended, has led to a number of problems in Western political philosophies and to hidden key paradoxes and problems, not the least of which is the tendency toward Platonic tyranny (pp. 114–129). Popper then addressed the importance of democratic institutions in a democracy. The point here is that, while there are innumerable philosophical and theological discussions regarding kinds of sovereignty, what is never questioned until the 19th century is the belief that sovereignty is existentially necessary for organizing society. I add here that the notion of sovereignty was also embedded in ancient narratives, such as *The Epic of Gilgamesh*, and religious texts. Andrew George (2003) argues that a key theme in this ancient story is the "eternal conflict between nurture and nature—articulated as the benefits of civilization over savagery" (p. xv).

45 It is necessary to point out that there is no such thing as *a* sovereign. The sovereign and sovereign classes exist in relation to apparatuses that carry out the sovereign's decisions (e.g., police, other law enforcement organizations, bureaucracies). The police are not part of the sovereign classes; they are functionaries of the sovereign. The sovereign classes include the political and economic elites who identify with and benefit from supporting the sovereign. As Chris Harman (2017) demonstrates, the rise of civilization and, concomitantly, of sovereignty accompanies the rise of sovereign classes and functionaries that are necessary for supporting the ruler and the exercise of state apparatuses or disciplinary regimes. For those who live in so-called democratic nations, it is clear that "democratic sovereignty" also has ruling and economic elites, despite the idea of political equality. This is especially evident when we note the vast disparities in income and wealth (Piketty, 2014, 2020), as well as the excessive political influence wielded by economic and political elites in constructing laws that benefit members of the sovereign classes (Gilens & Page, 2014).

46 While I will focus on a couple of paradoxes of sovereignty, it is worth mentioning that Wendy Brown (2010) identifies six paradoxes of sovereignty: (1) sovereignty is both a name for absolute power and a name for political freedom; (2) sovereignty generates order through subordination and freedom through autonomy; (3) sovereignty has no internal essence but rather is completely dependent and relational, even as it stands for autonomy, self-presence, and self-sufficiency; (4) sovereignty produces both internal hierarchy and external anarchy; (5) sovereignty is both a sign of the rule and the jurisdiction of law and supervenes the law; and (6) sovereignty is both generated and generative (pp. 53–54).

47 It is important to stress that violence is not always physical. Agamben (1998) writes that "it is the sovereign who, insofar as he decides on the state of exception, has the power to decide which life may be killed without commission of homicide. In the age of biopolitics this power becomes emancipated from the state of exception and transformed into the power to decide the point at which life ceases to be politically relevant" (p. 142). Yet, Agamben extends violence to include the power to make individuals and groups politically irrelevant, which is a central feature in the inverted totalitarian system present in the United States (see Wolin, 2008), as well as in indecent states that employ apparatuses of humiliation to produce and maintain subordinate and subjugated classes (Margalit, 1996). Individuals and groups cannot become politically irrelevant without state (and non-state) apparatuses and disciplinary regimes of humiliation that accompany the threat of physical violence. Nonphysical forms of political violence *violate* the political recognition of subordinate and subjugated others as persons, as citizens (see Soss et al., 2011; Wacquant, 2009). To violate individuals' recognition and status as persons (denial of self-esteem, self-respect, and self-confidence) is already a form of exclusion and political irrelevancy (marginalization from spaces of speaking and acting together), which is accompanied by material deprivation through failures in the adequate distribution of resources (Fraser & Honneth, 2003). The reality of food deserts, inadequate housing and housing insecurity, a dilapidated infrastructure, evictions, and the policing of poor persons and people of color are examples of social-political apparatuses of sovereignty that violate, threaten, and create political irrelevancy. The exercise of political violence, in these instances, may entail physical violence, but, more often than not, political violence is exercised through apparatuses that deprive people of rights, protections, and needed resources to survive and thrive.

48 There are also forms of political violence that are extrajudicial in the sense that the sovereign and sovereign classes tolerate or even promote this violence. For instance, Catherine MacKinnon (1991) notes that violence against women by men has been tolerated by sovereign states (patriarchal rule) for millennia. Similarly, in the United States, there is a long, sordid, racist history of terroristic lynching and raping of African Americans (see Alexander, 2010; Anderson, 2016; McGuire, 2011; Wilkerson, 2020). These acts occurred outside the juridical order yet were done with tacit approval of the white juridical order in that African Americans were denied justice. White supremacists were within the legal order but not subject to it during slavery and Jim Crow. In one sense, white supremacists decided (and decide) on the state of exception when they terrorized African American citizens. African Americans were and are included-excluded Others—within the juridical order but excluded from its protections. They could be killed, raped, or economically exploited, and these innumerable cases were not considered crimes. These acts of violence by white citizens were and are permitted by the sovereign and sovereign classes, even though these acts had no basis in law and were not directly perpetrated by the state.

49 There are other instances of exclusion or separation. In history, there is a process wherein human beings "began to detach man from his primordial animal condition" (Freud, 1927/1961a, p. 10). Exclusion is also noted in accepting Western hierarchal valuations of other species. Freud (1930/1961b) writes, "In the animal kingdom we hold to the view that the most highly developed species have proceeded from the lowest" (p. 68).

50 Interestingly, the occasional references to "savages" makes no appearance in the indexes of Freud's works (Freud, 1999). Freud (1913/1955e) uses the term, for instance, in *Totem and Taboo* (p. 99).

51 It is important to mention that Freud does refer to animal psychology and animals' dreams (1900/1953, pp. 131–132), animals' unconscious (1916/1957c, p. 189), and animals' thinking or cognition (1917/1955a, p. 140). But again, this is framed in terms of higher and lower, with human beings possessing "higher" capacities for thinking as compared with other sentient species possessing "lower" levels of cognition. In short, Freud's work retains Western philosophies' anthropocentric, hierarchal interpretive framework, which, like the term "primitive," has decidedly negative consequences for othered human beings and other species.

Chapter 2

Anarchic Care

Psychosocial Development and Dwelling in the World without the Rift

In the previous chapter, I mentioned Eduardo Kohn's (2013) claim that "all life is semiotic and all semiosis is alive" (p. 117). From simple amoebas to precocial bees to even more complex altricial human beings, semiosis is aimed at survival and flourishing. Another way to say this is that the semiosis of all living beings is aimed at dwelling in the world, and that existential insignificance represents the cessation of both semiosis and dwelling. The precarity of semiosis, of life, is the reality of existential insignificance. To narrow the focus, when Giorgio Agamben (1999) uses the metaphor of a burning house in reference to "the fundamental architectural problem [that] becomes visible for the first time" (p. 115), he is addressing how Western human beings' semiotic experiences and relations of dwelling in the world have, wittingly and mostly unwittingly, undermined the very material foundations for semiosis and dwelling for human beings and millions of other species—leading to an uninhabitable Earth (see Wilson, 2003). Western semiotic apparatuses have tragically contributed to the climate crisis and, in turn, exacerbated the precariousness of life, the precariousness of dwelling on and in this one Earth.

The proverbial canary in the coal mine in relation to the precariousness of the climate polycrisis and dwelling is the precipitous rise in eco-emotions among children and young adults. Scholar Panu Pihkala (2017a, 2017b, 2018a, 2018b, 2020a, 2020b, 2022a, 2022b, 2022c, 2024) has researched and written extensively about various eco-emotions, such as eco-anxiety, anger, grief, remorse, and fear. More particularly, Pihkala (2022a) points out that

> a recent global survey about climate change among 10,000 children and youth in 10 countries revealed that 56% of them thought that "humanity is doomed," while 75% felt the climate future to be frightening; 42% reported having felt at least some hesitation in having children because of the climate crisis.
>
> (p. 1)

This research clearly points to a wide range of negative emotions and cognitions. Terra Léger-Goodes et al. (2022) have researched the negative effects of

DOI: 10.4324/9781003518013-3

eco-anxiety on children, not the least of which is a loss of hope (see also Wu et al., 2020). Inmaculada Boluda-Verdú et al. (2022) explored the negative psychological and physical health effects of eco-anxiety and its relation to eco-depression, which are connected to feelings of helplessness and powerlessness. Closely associated with eco-anxiety is eco-grief, which has been researched by Cunsolo, Harper, et al. (2020; see also Comtesse et al. 2021; Cunsolo, Borish, et al., 2020; Cunsolo & Landman, 2017). The vast research on eco-emotions, in brief, reflects a crisis of dwelling that has its roots in the Western architectural flaw or ontological rift.

Another sign or symptom of this crisis is evident in the quest to leave the Earth to dwell on other worlds. Decades ago, when the Soviet Union launched *Sputnik*, Hannah Arendt (1958) was struck by a reporter's response. Instead of "pride and awe at the tremendousness of human power and mastery," there "was relief about the first 'step toward escape from men's imprisonment on the earth'" (p. 1). The Earth as a prison strikes me as not simply an expression of the alienation (and dissonance) that is spawned by the ontological rift; it is also fantastical in that escaping the material conditions for human life would mean paradoxically creating these very same conditions in space or on other planets. It is also fantastical in the unstated belief that life on other planets will support human life, and, even if another planet will support humans, why would we not also experience that as a prison? A more current sentiment is expressed in Elon Musk's and NASA's hope that human beings will become an interplanetary species (Taylor, 2023). This aim is connected to the human-caused degradation of this planet and the extinctions that will ensue. Again, we will have to create the material conditions to support human life (biodiversity) in order to live on other celestial orbs. Let's imagine this is possible. We will simply be colonizing another world and exporting the ontological rift with its destructive ripple effects. Nothing will have changed in how we dwell. In fantasy, nothing happens, nothing changes.

As daunting as the climate polycrisis is, can we imagine and construct semiotic apparatuses that reproduce neither the rift nor the attendant fantasies of escaping our imprisonment by living off-world? Can we render the apparatuses of the rift inoperative[1] and, in so doing, alter how we dwell with each other and other species? In the midst of the prevalence of the ontological rift, are there moments and relations where this rift is absent? More particularly, can we create a view of psychosocial development that illustrates the possibility of life without the architectural flaw? Providing one possible answer to this last question is the focus of this chapter. It is important to acknowledge at the outset that there are examples of Western human relations where the rift, while present in the societal apparatuses that shape our dwelling, is inoperative, though still functioning and negatively impacting othered human beings, other species, and the Earth. Inoperativity of the rift is evident in our friendships, in our love and care of pets, in nurses' care for dying patients, in persons' care for the flourishing of wildlife and habitats, and so on. While the apparatuses of the ontological rift are pervasive, they can be rendered inoperative through what I call anarchic caring relations, whether they are in relation to other human beings, other species, plants, and so on. More particularly, I argue in this chapter that good-enough parents' care for their infants represents a type of anarchic care. This anarchic care is foundational to infants' actualizing their potential for their initial semiotic dwelling in the world without the rift. Yet, as (Western) children develop, they eventually internalize the dominant societal apparatuses of the rift, and the early form of dwelling recedes, replaced by semiotic forms of dwelling informed by the ontological rift. I will depict this trajectory but, at the same time, argue for a developmental perspective that offers another possible viewpoint of dwelling that renders the rift inoperative, which we see glimpses of in Western life. Since parents' care is foundational for infants' experiences of survival and thriving, I begin with a definition of care before delving into a discussion of what is meant by anarchic parental care. From here, I focus on the psychosocial and psycho-relational development of infants and children, relying primarily on object relations and intersubjective theories as well as Western and Indigenous philosophical perspectives.

Let me offer some clarifications and caveats before starting. First, this chapter and indeed this book are explorations. This chapter does not present definitive universal views of development. My aim is to reimagine human development without the rift. Second, to construct an altered developmental perspective does not imply that the rift will disappear or that the perspective of dwelling in this chapter is *the* solution for the ills facing our world. It is rather an exercise meant to imaginatively amend not simply our theories but also our ways (including science) of dwelling in the world without the rift. In other words, my intent is to offer a developmental perspective that is ecological, acknowledging our absolute dependence on a biodiverse Earth. Third, while the language in this chapter tends toward universalization, I see this developmental perspective *as if* it is universal. In other words, I am telling a story about psychosocial development, and this story, I hope, has some resonance and relevance for people of other cultures.

Care, Its Attributes, and Anarchic Care

Since the 1980s, numerous philosophers have explored the notion of care and its relationship to politics, economics, and the environment (Bubeck, 1995; Engster, 2007; Engster & Hamington, 2015; Fraisart, 2021; Gilligan, 1982; Groys, 2022; Hamington, 2004, 2024; Held, 1995, 2006; Noddings, 1984; Oliner, & Oliner, 1995; Robinson, 1999, 2011; Segal, 2023; Sevenhuijsen, 1998; Tronto, 1993, 2013; Zúñiga, 2023). While it is understandable that definitions of care are contested in the academy, one nevertheless must make an attempt to define it. Broadly speaking, care is a semiotic activity aimed at the survival and flourishing of the objects of care. More particularly, for philosopher Daniel Engster (2007), care is "everything we do to help individuals meet their vital biological needs, develop or maintain their basic capabilities, and avoid or alleviate unnecessary or unwanted pain and suffering, so that they can survive, develop, and function in society" (p. 28). The strengths and limitations of Engster's view have been addressed elsewhere (LaMothe, 2017), but let's push ahead to a modified definition. Care is everything we do to help individuals, families, communities, and societies to (1) meet their vital biological, psychosocial, and existential or spiritual needs; (2) develop or maintain basic capabilities with the aim of human flourishing; (3) facilitate the inclusion and participation of other species[2] (and othered human beings) in the polis; and (4) maintain a habitable environment for all to dwell in (LaMothe, 2017, p. 8). I add to this definition that care, as a political concept, necessarily includes shared critical[3] and constructive reflection on how the structures (and their accompanying narratives and practices) of the state, governing authorities, and non-state organizations (e.g., businesses, labor unions, religious and secular communities) and actors meet or fail to meet the four features of this definition of care. I also stress that this definition of care has an ecological component that is foundational. Care requires tending to the habitat and its inhabitants, for our very dwelling in the world depends on a biodiverse Earth.

From this definition, I narrow the focus to the attributes of care as they pertain to the contexts of good-enough parents caring for their children. The first and foundational attribute of care is personal knowing (see Løgstrup, 1997; Macmurray, 1961), which is the recognition and treatment of children as unique or singular, valued, inviolable, and responsive or agentic subjects. For John Macmurray, personal knowing is both a matter of fact and a matter of intention. To the parent, the infant *is* a human-person, yet to recognize and treat the infant as a person indicates the parent's intentionality to recognize and treat this child as a singular human being. Negatively stated, we are all aware of instances of depersonalization[4] or failures to recognize infants (and adults) as persons, which lead to varied kinds of trauma.

As noted in the previous chapter, Macmurray also points out that all personal knowing is accompanied by object knowing. Simply stated, we locate objects in the world and seek to know them in terms of their various attributes (e.g., weight, color, size, texture, use). Object knowing, for Macmurray, appears first in development as infants seek to locate and engage objects in their environment. This early object knowing continues to develop throughout life. Cartesian–Baconian science, for instance, would be an example of a very sophisticated form of object knowing, though it is a kind of instrumental knowing that subordinates or eclipses personal knowing as it seeks to dissect or experiment on other species and, in some cases, human beings. Put another way, the Cartesian–Baconian sciences split off animistic knowing from object knowing, which means objects are mute and available for our use. In terms of the personal knowing that undergirds parental care, for Macmurray, object knowing must be subordinate to personal knowing if care has any chance of being effective and ethical. For instance, a baby cries, and the parent, recognizing the baby as a person, tries to evaluate the child's need or desire for the sake of addressing the baby's assertion. Parental attunement or mirroring comprises personal and object knowing, yet object knowing is subordinate to personal knowing.

Three other features of personal knowing need elaboration in relation to care. First, the sign "person" is, in part, empty. By this I mean that parents' personal knowing (care) creates a space for infants to appear in their singularity and, in appearing in their singularity, experience themselves as valued and unique, which I will develop further below. In other words, good-enough parental personalizing care facilitates infants' semiotic organizations of experiences of being-in-themselves and being-for-themselves. Second, personal knowing is an epistemological method of learning about the infant-person-object. Parents do not know fully their newborn infants (their alterity) and, in thousands of ways, as they care for them, they are learning about their children as these children seek to communicate. Put differently, the originally empty signifier "person" begins to take a particular shape given the singularity of infants and their assertions. This is also the case in therapy. I meet a patient for the first time, recognizing them as a person, and, in time, I learn about the various features of this singular person. Third, we typically associate the idea of "person" with human beings alone. For Indigenous peoples, personal knowing is a method that pertains to gaining knowledge about the habitat and its inhabitants (see Ingold, 2013; Kohn, 2013; Viveiros de Castro, 2017). Their infants are human persons. A crow is a crow-person. A bear is a bear-person, which means the bear's singularity is particular to this bear in this particular habitat where the bear and human dwell together. In brief, as an epistemological method, personal knowing is not necessarily aimed at human beings, and it is also foundational for Indigenous groups for dwelling together in this singular habitat and with its inhabitants. This will be important when I shift to infants' and children's psychosocial development.

Let me expand on the personal knowing of care as it relates to alterity because of its implications for infants' experiences of dwelling and our dwelling with other species. The very basis of care necessarily implies that the object of care is an other. Even if the infant is like the parent, it nevertheless is other-than-the-parent. The alterity of the other, in terms of care and personal knowing, is its singularity. *In caring, alterity is singularity.* Put another way, parental care creates a space wherein the singularities of infants are acknowledged and welcomed. When object knowing is separated from or split off from personal knowing, alterity lacks singularity, and we see distorted forms of care (e.g., deprivation of children, caring for animals who will undergo experimentation).

This singularity, whether in human beings or other species, can be further understood in terms of the idea that all life is semiotic and thus that each life is in-itself and for-itself in relation to its survival and flourishing.[5] Being-in-itself and being-for-itself do not fall under the heading of narcissism[6] but are simply semiotic features of all living beings that seek to survive and flourish in the face of the precarities of existence. Semiosis, which is necessary for survival and flourishing, accompanies senses of being-in-itself and being-for-itself. The implication here is that we share with all living beings the instinct or drive of being-for-itself, though for altricial animals, such as human beings, this is more complicated, as I will make clear.

Since I mentioned semiosis, it is important to state that parents' personal knowing and care are profoundly complex semiotic processes related to dwelling in the larger society. There are four points to be made here. First, I contend that parents' good-enough care is a semiotic activity that takes place in relation to the existential reality of insignificance. Parents realize the precarity of their infants' existence, their dwelling in the world. Without parental care, infants would cease to dwell, and this existential fact undergirds, in part, parental motivation to care—a motivation that *transcends* the social, political, and cultural meaning systems that shape care. Put differently, while parental care is shaped by the apparatuses of society, it is not determined by these apparatuses, which I will address in greater detail later. Second and relatedly, parents' personal recognition and treatment of infants as persons affirm their significance or value (being-in-itself and being-for-itself), which is over and against the dialectical reality of insignificance—the cessation of semiosis and dwelling. Third, as Winnicott (1971) noted, for parents to care for (hold and handle) their children, the larger society must recognize and treat parents as persons. As a result, parents' experiences of singularity and dwelling in the world are affirmed and supported. This seems obvious, but there are innumerable examples wherein the inhabitants of the larger society fail to recognize and treat parents as persons (e.g., racism, sexism, classism), leading to their struggles to dwell, and to provide their children with experiences of being at home, in the world. Fourth, while shaped by the apparatuses of the larger society, parental care may, at times, render these apparatuses inoperative. Let me return to Ruby Sales's experiences of care mentioned in the previous chapter. Ruby grew up in the Jim Crow South, where the apparatuses of white supremacy (e.g., depersonalization) undermined the dwelling of innumerable African American adults and their children. Ruby's parents personally recognized and treated their children as persons, even while the parents faced the depersonalization of White supremacy that undermined their experiences of dwelling and belonging in the larger society. Put another way, their care rendered inoperative the society's apparatuses of white supremacy by creating spaces for their children to experience being at home in the world and society even while this very same world sought to undermine their experiences of dwelling.

As to the last point, this is where we note the presence and importance of anarchic care. Before explicating what I mean by anarchic care, it is necessary to take a moment to say a few words about the highly evocative term "anarchy." The notion of anarchy appeared in the English language in the 17th century, though there were hints of it earlier in Western thought (see Prichard, 2022; Sheehan, 2003). The term simply means without a ruler. The very idea of the absence of a sovereign ruler for organizing political life must have nauseated and terrified Thomas Hobbes (and his contemporary John Milton), who railed against the idea of anarchy and thought freedom and societal stability depended on submitting to a powerful leviathan (Ryan, 2012, p. 297). Anarchy, Hobbes believed, led only to political-social chaos and violence—a society where life would be brutal and short. Not surprisingly, Hobbes, given his own privileged social position, seemed to ignore the overt and hidden forms of violence perpetrated by sovereign classes and their apparatuses, yet Jean-Jacques Rousseau (2016) seemed to recognize this in the last lines of his *Discourse on Inequality*: "It is plainly contrary to the law of nature … that the privileged few should gorge themselves with superfluities, while the starving multitude are in want of the bare necessities of life" (p. 126). In the 19th century, Pierre-Joseph Proudhon and others sought to resurrect the idea of anarchy or non-sovereignty (Newman, 2016, pp. 11–12), yet today anarchy retains strong negative valences that, in my view, function to foreclose our ability to imagine political life or dwelling without some form of sovereignty—a feature of the rift.

To return to the idea of anarchic care, in the previous chapter's discussion of the ontological rift I indicated that a central feature of this rift is the apparatuses that produce and maintain dominion (sovereignty) over othered human beings and other species and are accompanied by beliefs in the superiority of those exercising dominion. As I noted, sovereignty, especially as it relates to othered human beings and other species, attends instrumental (not personal) epistemologies and negative power—power that aims to actualize the other's potentiality as mere use value. In other words, the ontological rift produces forms and gradations of depersonalization that deny the singularities of othered human beings and other species and affirm their mere use value. Given this, I contend that the earliest forms of parental care are anarchic. By this I mean that parents, in recognizing and caring for their infants, render inoperative those apparatuses of the ontological rift. Good-enough parental care does not accompany being sovereign over their infants or a belief in parents being superior to infants. To be sure, parents obviously have more power, authority, and responsibility than their infants, but this asymmetry is not the hierarchal power of sovereignty's dominion over others that is noted in Laius's treatment of his newborn son and Jocasta's silence. Stated positively, anarchic care is the exercise of positive power that entails the personal recognition of, respect for, and treatment of infants' singularities.[7] Anarchic care, then, creates a relation wherein altricial infants begin to actualize their potentialities for varied experiences of singularity and dwelling/belonging.[8]

Of course, anarchic care does not mean the apparatuses of the ontological rift no longer operate and have negative effects. Rather, for Ruby's parents, apparatuses of dominion (white supremacy) were rendered inoperative through their personalizing care but nevertheless impacted them. The parents' anarchic care rendered these apparatuses inoperative, thus largely, but not completely, shielding their children from the negative power of racism's depersonalization. More positively, the parents' anarchic care created spaces for the children to experience their singularity in mutual personal relations—in dwelling together. As Ruby notes, as a child she felt like a first-class human being even in a society dominated by the apparatuses of white male sovereignty and supremacy.

As a brief aside, anarchic care is not simply evident in parent–infant relations. Aristotle (1971, p. 180) and, more recently, Macmurray (1961, p. 151) point out that friendship entails a mutual, personal, anarchic caring relationship wherein each is treated as an end and not as a means. Each person affirms, safeguards, and delights in the alterity and singularity of the other. There are other routine examples of anarchic care in adult life, such as a nurse caring for a dying patient, a therapist listening to a patient's story, a woman opening the door for an elderly man, a man recognizing the singularity of and thus caring for his cat, or a person helping out a neighbor. The complexity of life in the rift is reflected in these acts of anarchic care coupled with the presence of the ontological rift and its negative effects. For instance, all of these illustrations of anarchic care can coexist with the ontological rift within each person. After leaving a friend, one goes to the grocery store to purchase meat from industrial slaughter factories, or to the pharmacy to obtain medications that were developed by drug companies experimenting on animals or beauty products that are developed by companies using animals for testing products. We are indeed contradictory and, for those of the rift, tragic creatures.

There are three additional features of parental attunement or anarchic care to tease out. Above I noted that all life is semiotic and all living beings manifest the drives of being-in-itself and being-for-itself. Good-enough parental personalizing care can be understood as comprising the additional aims of confirming an infant's drive to and experience of being-for-itself *and* being-for-ourselves (parent–infant dyad). Put differently, parents' recognition of infants' singularities accompanies a categorical command to care, which entails parents' setting aside (not getting rid of) their drive of being-for-themselves and placing in the foreground the drive of being-for-the-other. As Danish philosopher K. E. Løgstrup (1997) argued, the "ethical demand takes its content from the unshakable fact that the existence of human beings is intertwined with each other in a way that demands human beings protect the lives of others who have been placed in their trust" (p. 290). Parental anarchic care, then, represents at times the ability to set aside (not get rid of) a parent's being-for-itself for the sake of creating a relation and space for infants to organize experiences of being-in-itself and being-for-itself (infants' singularities). This is quite abstract, but consider a parent, who is exhausted from work and parental responsibilities, getting up to care for their sick child at 3 a.m. The parent sets aside their own needs and desires to create a space to care for the well-being of the child. Care, in this case, is not the parent being-for-itself, at least not in the foreground, but the parent being-for-the-other (the child).

A second and related feature of anarchic care is being-for-ourselves—parents and infants' dwelling together. While I am suggesting that all human beings (and other species) manifest the drive of being-in-itself and being-for-itself, parents' personalizing attunements include the motivations of being-for-the-other *and* being-for-and-with-ourselves (infants, parents, and other family members—and, later, others of the larger social group). By this I mean that, as social and communal animals, there is a motivation to engage in reciprocal, cooperative, and intimate relations with, in this instance, infants, even given that this is mostly potential and not actual in infants. For instance, Erik Erikson (1982) conceived the first stage of psychosocial development in terms of a "dialogue" between parent and infant, wherein the parents' "almost unrestricted attentiveness and generosity" give rise to the child's basic trust (p. 35). Other infant–parent researchers have noted that parents engage in proto-conversations with their infants (Bonovitz & Harlem, 2018; Levin & Trevarthen, 2000; Trevarthen, 1993). They act *as if* infants can understand and engage in cooperative, reciprocal intimate dialogues. Proto-conversations represent, in my view, the drive of being for and with others—being-for-ourselves.

Two other aspects of dwelling and of personalizing anarchic care with regard to human psychosocial development are potentiality and actuality. For Agamben (1999), all living beings, depending on material and other factors, actualize their potentialities. Infants are wonderful examples of the relation between potentiality and actuality. Infants are potentially capable of using symbols, language, self-reflection, political agency, and so on. A question that Agamben does not answer is, what is necessary to facilitate, in this case, infants' actualizing their potentialities? I contend that parents' reliable personal or anarchic caring provides a relation and a process through which infants begin to actualize their potentialities, which I will develop below. For now, it is important to add that Agamben differentiates the potentialities of human beings from other species, though he does so without reproducing the ontological rift. This differentiation is important for our discussion of anarchic care. Agamben argues that "the very essence of humanity lies in a potentiality that is expressed when it does not unfold into actuality" (Colebrook & Maxwell, 2016, p. 289). Here is where Agamben (1999) turns to illustrate potentiality in terms of impotentiality. He wrote:

> *Other living beings are capable only of their specific potentiality; they can only do this or that. But human beings are the animals who are capable of their own impotentiality. The greatness of human potentiality is measured by the abyss of human impotentiality.* Here it is possible to see how the

> root of freedom is to be found in the abyss of potentiality. To be free is not simply to have the power to do this or that thing, nor is it simply to have the power to refuse to do this or that thing. To be free is … *to be capable of one's own impotentiality.*
>
> (pp. 182–183, emphasis in original)[9]

Agamben illustrates this by relying on Herman Melville's *Bartleby, the Scrivener*, wherein Bartleby was asked by his boss to do something and Bartleby replied, "I prefer not to." For Agamben, Bartleby was exercising the freedom of not actualizing his potentiality, which means, in part, that Bartleby was not determined and could not be determined (in the sense of being commanded by others) to actualize his potentiality. Colebrook and Maxwell (2016) add that "to have potentiality is to be capable of not becoming what one has the capacity to be" (p. 38). A concert pianist, for instance, can prefer not to actualize their capacity to perform. Human beings, then, are not (and cannot be completely) compelled to actualize their potentiality because they possess the agency and freedom to prefer not to.

The notions of potentiality and impotentiality are closely associated with Agamben's depiction of "inoperativity," which means deactivating the function of the apparatuses. It is important to stress that inoperativity, for Agamben, is not passive (Prozorov, 2014, p. 134). That is, inoperativity does not "affirm inertia, inactivity or apraxia … but [is] a form of praxis" (p. 33). Bartleby's impotentiality, for instance, renders inoperative the apparatuses associated with his job. This said, the exercise of impotentiality does not mean that these apparatuses cease to operate or have effects (pp. 31–34). Put another way, individuals who enact inoperativity are not captive to the grammar of the apparatuses, even if these apparatuses continue to have their effects. So, Bartleby, for example, exercises his impotentiality by preferring not to fulfill his boss's demands, thereby rendering inoperative the apparatuses of, in this case, capitalism. Of course, Bartleby's act of impotentiality does not diminish the negative power of capitalistic apparatuses to impact him.

Now, what does all of this have to do with anarchic care? Good-enough parents' care for their infants renders inoperative the apparatuses of sovereignty (dominion over) and beliefs in superiority and inferiority—apparatuses of the ontological rift. Interestingly, Freud used the Oedipal story to frame psychosocial development. In this story, Laius (and Jocasta) did not care for his son. Laius operated out of apparatuses of sovereignty and its negative hierarchical power to foreclose his son's actualizing his potentialities (including impotentiality) by ordering him to be killed. Jocasta's silence represents the absence of care and an inability or unwillingness to render inoperative the apparatuses of sovereignty for the sake of her son. Some parents may seek to have dominion over their children and believe themselves to be superior, but I think these are destructive relations, as evidenced by the story of Laius and by Alice Miller's (2002) research into 19th-century childrearing practices in

Germany. More positively stated, parents' anarchic care reveals their *preferring not to* (inoperativity) actualize apparatuses of the rift, which represents a type of positive power that creates spaces for children to actualize their potentialities, including their potentiality for impotentiality.

What we have in common with other species is that all life seeks to dwell in the world—to survive and flourish. Similarly, all life has a semiotic sense of being-in-itself and being-for-itself, though what this looks like is infinitely variable. As far as we know, what distinguishes us from many species (not all) is that we are altricial, social-communal animals. This means we require the care of others to actualize our potentialities. What further differentiate us are our capacities for symbolization, impotentiality, self-reflection, complex reasoning, and self-consciousness.[10] All of this impacts our ways of caring for others and our dwelling in the world.

In this section, I have argued that personal-object, anarchic knowing is a foundational feature of good-enough parental care. This good-enough anarchic care (1) represents parents' setting aside their being-for-themselves, (2) represents parents' being-for-others (their infants) and being-for-ourselves (being-for-and-with-others), and (3) renders inoperative the apparatuses of the ontological rift, all of which create spaces for infants to actualize their potentialities for being-in-itself and being-for-itself.

Reimaging Psychosocial Development without the Rift

We can gain a clearer picture of psychosocial development and dwelling by shifting our focus to infants and children. In this section, I begin with prebirth organizations of experience, arguing that the semiotic drives for being-in-itself and being-for-itself are evident in the organizations of experience manifesting infants' preferences. From here, I depict the first experience of being unhoused—birth—and the emergence of nascent consciousness and communication. Relying on the amended work of Donald Winnicott, I depict infants' first use of an object—the primary transitional object (PTO)—which attends an early semiotic epistemology of object knowing and animism. As infants begin to actualize their capacities for symbolization, a secondary transitional object (STO) emerges, representing more complex organizations of experiences, relations, and dwelling. This STO builds on earlier epistemologies of object knowing and animism, revealing early epistemological principles of multinaturalism, perspectivism, and personalization. This leads to a discussion about how these early forms of dwelling and epistemologies are, in the West, largely replaced or subverted by the internalization of the ontological rift. I then argue for the notion of tertiary objects, which replace STOs. These tertiary objects can be understood as operating in two directions. One direction involves the internalization of the rift and consequent splitting-off of earlier ways of knowing (e.g., animism). The other direction involves integrating earlier epistemologies into more complex epistemologies that include the personalization of other species. The last section, then, depicts later development that is not captive to the rift—one that includes recognizing, respecting, and caring for the singularities of other species as well as for the flora and fauna of the habitats in which we dwell.

Infant–parent researchers, over the last four decades, have shown that a baby develops preferences prior to birth (Beebe & Lachmann, 1994, 2002; DeCasper & Fifer, 1980; DeCasper & Spence, 1986; Kumin, 1996). This suggests that psychic-semiotic (pre-representational or nonpropositional) organizations of experience exist prior to birth. The womb is the material condition for these earliest organizations of embodied experiences of dwelling. In other words, the earliest pre-representational, embodied senses of being-in-itself and being-for-itself depend on the apt material conditions of the womb. Moreover, the idea that infants, in utero, are organizing experience in terms of preferences suggests two other features of early development, namely, the presence of nascent agency and that of resonant relatedness. Decades ago, Winnicott (1975) surmised that "birth can easily be felt by the infant" and that they participate through "personal effort" (p. 186). Here we see the notion of the infant's pre-representational "belief" that the infant participates in the birth, which implies the presence of agency (an aspect of the ego). In short, the organization of embodied, pre-representational experiences of dwelling points to a nascent bodily agency—an agency necessary for dwelling. As Matt Waggoner (2018) notes in his discussion of Theodor Adorno's philosophy, "The human experience of dwelling begins with embodiment, with the fact that [nascent] consciousness is inseparable from the somatic and sensorial" (p. 105). In addition, pre-birth organizations of preferences suggest that nascent agency is relational. This can be understood from an evolutionary perspective. Infants' semiotic pre-representational organizations of preferences are inextricably tied to the drives to survive and flourish. To survive and thrive depend on an agency directed toward the source of the infants' physical and psychological survival, namely, the parents. In brief, dwelling in this first home is an embodied, pre-representational, semiotic organization of being-in-itself and being-for-itself that is joined to nascent relational agency.

There is one other feature of this early existence and semiosis. Hartmut Rosa (2019, 2020) develops the term "resonance,"[11] briefly arguing that it "is constitutive not only of human psychology and sociality but also of our very corporeality, of ways we interact with the world tactilely, metabolically, emotionally, and cognitively" (2020, p. 31).[12]Rosa (2019) stresses that "*resonance is not an emotional state, but a mode of relation* … a specific way of *being-related-to-the-world*" (pp. 168–169, emphasis in original). This "basic mode of vibrant human existence consists not in exerting control over things but in resonating with them" (2020, p. 31).[13] While Rosa (2019) is not focusing on pre-birth resonance,[14] the term is fitting here since he believes resonance is "a basic human capacity and need" (p. 170).[15] Prior to birth, infants' dwelling and semiotically organized preferences are resonant with the parental womb. Rosa writes that "resonance in its full sense occurs only when … we feel connected to the world" (p. 33), and this connection is understood at this stage as embodied dwelling and belonging.[16] Put another way, early semiotic constructions represent the embodied, pre-representational resonance between the infant and the parental womb.[17] This resonance is contingent on the parent's recognition of and respect for the infant's singularities, which means that there must be some acceptance of their inability to control the infant. For Rosa (2020), the very possibility of resonance depends on surrendering to the fact of the uncontrollability of the singular other. Good-enough parents' recognition of their children's singularities means there is, amid their anarchic, agentic care, a surrender to uncontrollability. The very possibility of resonant dwelling, then, depends on semiosis, singularity, and some measure and acceptance of uncontrollability.[18]

Naturally, birth disrupts these earliest embodied, presymbolic organizations of resonant dwelling. Initially ensconced in the warmth and dampened quiet of the womb, the newborn is thrust into a sensorially complex world—"one great blooming buzzing confusion" (James, 1918/1956, p. 488). At birth, infants are, if you will, unhoused,[19] which means that there is not only much that is new but also a gap between need and the meeting of that need. Moreover, this gap accompanies a disruption in infants' semiotic embodied resonances, which initiates a semiotic impulse to communicate. Previously, in the womb, there is no gap between need and the meeting of the need, which may be a reason why some developmentalists associate the earliest period of development with Eden-like experiences where all needs are met before they can rise to consciousness. Once the infant is born, their needs begin to be experienced because of the gap in time between the need and meeting of the need. This gap gives rise to the emerging consciousness, the semiotic impulse to communicate, and developing agency. John Macmurray (1961) writes that the infant "is, in fact, 'adapted', so to speak paradoxically, to being unadapted, 'adapted to complete dependence.' […] He can only live through other people" (pp. 8, 51). In infants' unadapted, dependent, and vulnerable state, they possess an impulse or motivation to communicate—"the impulse to communicate is [their] sole adaptation to the world" (p. 60). That is, to survive and thrive, to dwell resonantly, the altricial infant must communicate and cooperate with the sources of nurture. As indicated above, pre-birth semiotic organizations of preferences for the mother's voice are the preparatory ground for infants' impulse to communicate.

Beginning with Freud, psychoanalysts recognized the importance of optimal frustrations for development. These frustrations (and attendant aggression) emerge because needs are not met instantaneously, which is a necessary feature for engaging in reality rather than fantasy. Winnicott (1971) argued that, in the earliest period of development, the parent is preoccupied with meeting the needs of their infant, which provides the infant with the experience of and belief in their own omnipotence.[20] Eventually, as the infant develops, maternal preoccupation wanes, which sets up infants' necessary experiences of disillusionment that are necessary for recognizing and responding to reality. Another way to frame this is that the inevitable frustrations experienced by infants, because of the gap between experiencing a need and the need being met, concern the emerging evolutionary, necessary connection between being-in-itself, being-for-itself, and *being-for-and-with-ourselves*. That is, the assertions of needs and the parents' meeting those needs in a timely manner reflect the early cooperative, resonant communication that is a necessary foundation for social-communal life—for shared survival and thriving—and for a resonant sense of being alive together. The reality principle, in other words, concerns the question, how are we to dwell pragmatically and resonantly in this world together?

Infants' impulse to communicate is met by parents' personalizing anarchic care (attunement, mirroring) that recognizes and affirms infants' singularities. These good-enough parent–infant cooperative interactions, I suggest, contribute to emerging semiotic, embodied organizations of self-esteem, self-confidence, and self-respect that are integral for infants developing agencies and experiences of dwelling in this new, complex world.[21] In other words, the pre-birth embodied, pre-representational organizations of being-in-itself and being-for-itself are altered by parents' reliable personalizing care, taking shape as senses of relational, embodied self-esteem, self-respect, and self-confidence, which are foundational for emerging social-political agency.

It is important to pause and say a bit more about social-political agency. As mentioned above, infants possess numerous potentialities that, in time, are actualized. Infants are born into a polis, and, while they have nascent agency, their political senses of self and agency are only potential. Andrew Samuels (1993, 2001, 2004, 2015) has long argued for the importance of attending to the development of political selves in psychoanalytic theorizing and therapy. For Agamben, the political self, nascent or otherwise, is "at the root of both politics and ethics" (Kotsko, 2020, p. 60). Put another way, political agency is about dwelling together, whether we are addressing parent–child relations or the larger polis. In terms of the ontological rift, political agency and dwelling and belonging have excluded other species, yet this is not yet an issue with regard to infants' nascent or potential political selves and agencies. Nor is it an issue regarding parental care. Good-enough parents possess political self and agency and are shaped by the apparatuses of the polis, yet their caring attunements include rendering these political apparatuses inoperative when they are seen to obstruct or hinder care.

In terms of infants developing a sense of self and agency, no rift exists between themselves and other animate and inanimate objects. Stated differently, infants' early form of knowing and engaging the world comprises animate object knowing that is devoid of beliefs in superiority and inferiority. This pre-political world of objects is alive, which is the foundation of the impulse to communicate and a method of knowing. That is, animistic knowing is a method of coming to know and relate to a world of multiple singular objects—objects that are believed to be alive as the infant is alive. Put differently, the objects of the world are alive and, in being alive, agentic. This early nonpropositional knowing affirms multiple and varied agencies (Latour, 1996, p. 5). If the ontological rift is actualized, it happens only later in development, as Western children internalize the political apparatuses that produce the rift.

It is important to stress that the presymbolic sense or belief that the world and its objects are alive is inextricably a part of the self and infants' senses of resonance with the world. The child's sense of aliveness is connected to the aliveness of the world, and this is accompanied by a sense of realness or truth. Put another way, this is a living world of belongingness, of resonant dwelling together. I will say more about this later, but the eventual internalization of the apparatuses of the rift (for Western subjects) means that the world subjects encounter is mute and dumb. This results in a loss of our self being alive in a world that is alive (Rosa, 2020, p. 22). We lose a sense our sense of resonance because "what appears to us must be known, mastered, conquered, made useful" (p. 6).[22] But I am getting ahead of myself.

Winnicott's (1971) notion of transitional objects and phenomena can help in further depicting psychosocial development and experiences of dwelling. To understand the child's transition from relatively undifferentiated, embodied relations to engaging and using objects, Winnicott posits that a transitional object (TO) in early childhood represents the first not-me possession, which entails the infant's rudimentary ability to recognize an (alive) object as independent of the infant (e.g., an external object, pp. 3–4). In the act of possessing and using this object, the infant, Winnicott claims, both retains and partially hands over their belief in and experience of omnipotence (pp. 9–11). There is, for Winnicott, a paradoxical interplay of the internal and external reality. This PTO[23] is "not an internal object—it is a possession. Yet, it is not (for the infant) an external object either" (p. 9). Strictly speaking, the TO, for Winnicott, is neither an internal object nor an external object, and yet it is both. Another related feature of TOs is that they must not be changed unless by the infant because this would challenge the infant's sense of and belief in their omnipotent control and, consequently, disrupt their experience of continuity or sense of going on being (p. 4)—dwelling as embodied-being-with-other. Similarly, the parent is not to challenge the child's omnipotent selection and use of the PTO because this too would challenge the child's *presymbolic* belief in and experience of omnipotence, which Winnicott contends is needed for the child to have confidence in organizing experience.

Since the PTO is not simply identified with either the external or internal world, a question is raised regarding the child's selection of the object and what the TO represents to the child. The object, which may be presented to the infant, is nevertheless chosen by the infant and not shared.[24] This choice of a *primary* TO is in "accordance with its consistency, texture, size, volume, shape, and odor" (Kestenberg & Weinstein, 1978, p. 89), which is psychologically joined to the "technique of mothering"[25]—the caregiver's handling, holding, comforting, and consoling the infant (Winnicott, 1971, p. 11). That is, the child unconsciously chooses a PTO that represents the parent's anarchic care for the child and attendant embodied experiences of resonant dwelling, which permits the child to manage (regulate) their emotions during times of anxiety or separation from the parent.[26]

All of this can be summarized in terms of the experience of dwelling. This first object under the child's "omnipotent" control (belief) provides embodied soothing, rest, and continuity—experiences of being-at-home-with-an-object—an object that is me and not-me. In other words, *being-in-itself and being-for-itself are joined to being-with-an-other.* This "other" is not recognized as a person, but it is, for the infant, animate—a living object that dwells. There are four important points here. First, this is an early form of animate knowing and relating to an object, one that is believed to be alive. Second and relatedly, this early epistemology, if you will, continues to develop, becoming more sophisticated or complex but nevertheless present in later forms of knowing.[27] More precisely, this early method of animate knowing is foundational for engaging in and dwelling with others and other species, which will become clearer as we proceed. Third, there is no ontological rift manifested in this form of knowing and relating with the PTO. Fourth, it is important to explain that this early embodied experience is not to be understood as if one is housed alone in one's body (in utero being-in-itself and being-for-itself). In discussing Adorno's work on dwelling, Waggoner (2018) writes that "no one ever singularly inhabits a body because embodiment is not … a state of being, a self-sufficient thing. Its existence is inseparable from and can only be constituted as such within a matrix of contact and connectivity" (p. 107). That is, "The [nascent ego] is housed by extended realities that are material and social in nature" (p. 109). So, from the very first, embodied dwelling is relational—embodied dwelling is resonant belonging.

As infants develop, more complex symbolic capacities emerge, as do relations with other animate and inanimate objects. I differentiate between the PTO and the STO because there are significant differences between the proverbial blanket and the child's use of cultural objects as TOs. STOs represent the emergence and ongoing actualization of symbolic semiotic capacities (e.g., language, narrative, self-reflection). Perhaps one could say that a fundamental question—and task—regarding STOs is how to dwell with others qua persons.

Because Winnicott elevated the importance of play in human life, I use the comic strip *Calvin and Hobbes* (Watterson & Trudeau, 1985–1995) to illustrate my claims regarding STOs. Calvin is a little boy who plays with his stuffed tiger, Hobbes—a cultural object. Together they create and inhabit a world together. Hobbes and Calvin embark on all kinds of adventures; they argue and make up and they comfort each other when hurt or distressed. Calvin's parents are present but in the background, and, when they do appear,

Hobbes, from the parents' perspectives, is a lifeless object. To Calvin, however, Hobbes is very much alive, and not just alive but a *tiger-person* who, Calvin believes, recognizes and treats Calvin as a human-person. In Calvin's imagination, there is, then, a sense of mutual self-esteem, self-confidence, and self-respect that is derived from speaking and acting together, which includes shared repairs of conflict. Although it is in Calvin's imagination, they together, in speaking and acting together, create a pre-political space of appearances wherein they reside.[28] As an STO, Hobbes serves as a process toward Calvin's learning to dwell resonantly with others—being-for-and-with-ourselves. It is crucial to stress that the "others" are not simply human-persons but other living beings. Hobbes, in other words, represents the possibility of recognizing and treating other living beings as persons.

Let me address this in another way. In my view, STOs do not represent the challenge between inner and external reality but rather a lifelong relational challenge of dwelling, namely, how to navigate being-in-itself and being-for-itself with being-for-and-with-others (other living beings, not just human beings). Calvin, for instance, has symbolically organized experiences of being-in-himself, being-for-himself, and being-for-and-with-others in relation to Hobbes, who Calvin believes has similar capacities and experiences. The challenge is how these two singular beings manage to mutually care (being-for-others) and cooperate such that they can dwell resonantly together. While this is clearly imaginative play,[29] it nevertheless, in my view, represents a key epistemological factor with regard to dwelling. In terms of the STO, there is no ontological rift. Hobbes is known or recognized as a tiger-person and treated as a person who dwells with Calvin. There is no epistemology with imbedded beliefs in superiority and inferiority between them. Dominion does not exist in their relationship or in Calvin's animistic knowing, which now entails personal recognition.

In adult life, we see glimpses of this form of knowing and dwelling. Many people who live with their dogs and cats recognize and treat them as persons. That is, they recognize that these animals have personalities and are unique. They do not confuse cats for human persons but recognize their cats as cat-persons and, in so doing, share in their dwelling resonantly together. Similarly, many Indigenous peoples recognize other species as persons. A beaver is a beaver-person. A bear is a bear-person. I wish to stress that personal epistemology is not simply playful or imaginary. It represents an evolutionary epistemological method, which is evident in learning to dwell with other species. Put another way, this animistic personalizing epistemology represents a method of learning about the singularities of others that inhabit the world with us.[30] This epistemology is necessary for the survival and flourishing—not just of human beings but of other species that dwell with human beings in this particular habitat. Calvin, we can imagine, is learning to cooperate with other-than-human animals for the sake of reciprocal, resonant dwelling.

Some may object, pointing out that children's playful imagination is radically different from adults' more sophisticated, rational, and ostensibly realistic approaches to dwelling. Granted, children continue to develop and learn more complex ways of knowing and engaging the world. However, the objection also reveals the presence of the ontological rift. What happens to Hobbes, according to Winnicottian thinking, is that he is not mourned, having lost significance as Calvin learns to socialize and play with others. Yet, I suggest that in the West children begin to internalize the apparatuses of the ontological rift, which include beliefs in the superiority of reason and science, the necessity of dominion and mastery for dwelling, and the putative inferiority of animistic epistemologies. Put another way, animistic thinking is infantilized and seen as something to overcome. Of course, Freud believed this early form of knowing is always present in adult life and dwelling, but when operative he pathologized it. Yet, consider the popularity among adults of the comic strip *Calvin and Hobbes*, the *Toy Story* films, or love of pets. One may argue that this is simply a nostalgic form of adult play, which may be true, but it is not the whole story.

Let me take a short detour before returning to psychosocial development and dwelling. In Chapter 1, I mentioned anthropologists who have sought to decolonize Western anthropologies with the aim of better understanding Indigenous peoples. Studying various Indigenous peoples in Central and South America, Eduardo Kohn (2013), Tim Ingold (2013), and Eduardo Viveiros de Castro (2017) note that these peoples view other species as persons or potential persons, and that this personification, if you will, is necessary for gaining knowledge of the other creature, the habitat, and themselves. This does not imply that Indigenous people believe that other species qua persons are identical to human beings qua persons. As stated above, they are crow-persons, bear-persons, lion-persons, and so on. In this epistemology there is a multiplicity of perspectives as well as multinaturalism.[31] Animistic personification or epistemology, then, means recognizing and respecting the singularities of other species, which include the recognition of the multiple natures and agencies in the habitat. That is, Indigenous peoples come to know other species in their singularities, which implies the presence of the epistemological principles of multinaturalism and perspectivism. Moreover, this form of knowing, for Indigenous peoples, is necessary for resonant living and dwelling in and with nature. Rather than trying to control, master, or dominate "Nature," they see themselves as intrinsically a part of "nature" and thus learn to adapt to and cooperate with nature. Put another way, regarding other species as persons or potential persons means that they are considered an inextricable part of *a common world of resonant belonging*. The polis of Indigenous peoples includes other species and the habitat. The fact that Indigenous peoples have lived and dwelled sustainably for millennia provides evidence of the practicality and ecological effectiveness of animistic epistemologies that personify other species.

Critics may point out that this form of knowing is not scientific. That would be correct if the "sciences" were limited to the Cartesian–Baconian perspective, which objectifies and seeks dominion over nature and other species to the exclusive benefit of (some) human beings. Yet, Indigenous peoples observe, test, and analyze in relation to their environment and its inhabitants. To be sure, their interpretations may be illusory,[32] but the aims of their personalizing, animistic epistemologies are survival and flourishing, which necessarily include practical or realistic adaptation to and cooperation with other beings that dwell with them in the habitat upon which all their lives depend. I add here that not all Western science is captive to the ontological rift. There are scientists who recognize and respect the singularities of other living beings they are investigating (see Goodall, 2010; Margulis, 1997, 2007; Montgomery, 2016; Schlanger, 2024; Yong, 2023). They may not construct members of other species as persons, but they treat them as fellow creatures by recognizing and respecting their singularities, thus eschewing the Cartesian–Baconian scientific approach of objectification and dominion that leads to ecological dissonance.

To return to psychosocial development, is it possible to imagine children's development that does not lead to the infantilization and repression of personalizing, animistic thought but instead fosters the principles of multinaturalism and perspectivism? Are children's subjectivities in the West determined by the apparatuses of the rift—apparatuses that have clearly undermined the dwelling of human beings and millions of other species? Can we find ways to extend anarchic care to other species? In other words, can we, as adults, render inoperative the apparatuses of the rift so that we might recognize the singularities of other species, as well as recognize that all species, sharing the precarity of existential insignificance, seek to survive and thrive—dwell—in and on the Earth? There are, of course, numerous examples of adults who are not captive to the rift, whether these are scientists or lay persons. Authors of children's literature use personalizing, animistic thought to convey not simply morals or values but ways of dwelling with other species. Some parents encourage their children to imagine the singular life of another animal or even of trees when they go on hikes. Of course, this is over and against the massive systemic apparatuses (exemplified by the Cartesian–Baconian sciences, capitalism, petrocultures, imperialism) that produce relations of domination and depersonalize not only other species but also many human beings. Nevertheless, there are possibilities of human dwelling where personalizing epistemologies include other species and habitats. In part, these possibilities depend on psychosocial development that renders the rift inoperative, which creates spaces for children and adults to dwell with other species on this one Earth we all share.

Back to Calvin, psychosocial development, and Rosa's (2020) notion of resonance and its attributes. The first attribute of resonance is *being affected*, which is to be moved by the other; in being moved, affectively and physically, one experiences a sense of being addressed (p. 32). This follows my depiction of mutual personal recognition. One cannot be moved in this way without the presence of trust and care. Calvin, in his imagination, is moved by Hobbes, and he believes Hobbes is moved as well. In Calvin recognizing Hobbes as a tiger-person, and Hobbes, in Calvin's imagination, recognizing Calvin as a human-person, each addresses and experiences being addressed by the other. The second attribute of resonance is self-efficacy, which is not only the sense of being connected but also the experience of having an effect on the other and the world. Calvin and Hobbes play, argue, and so on, manifesting their shared sense of having an effect on each other and the environment. The third attribute of resonance is adaptive transformation, wherein we feel alive with others. Rosa contends that "resonant relationships … change both subjects" (p. 35). Mutual anarchic care and creativity characterize how each affects the other and how each is changed. Finally, resonance "is inherently uncontrollable" (p. 37). This may at first appear not to fit with Calvin's imaginative play since he is ostensibly in control. I would argue that, in believing that Hobbes is a person, Calvin is already creating a space not to be in control. To be in omnipotent control is to deny the singularity of the other and to undermine creativity and change. Calvin believes Hobbes possesses his own mind, agency, and so on, and that he is, therefore, uncontrollable. To echo Winnicott (1971), paradox must be tolerated and accepted (p. xii).

Imagine that Calvin, having learned to engage in reciprocal, resonant, personalizing, and cooperative relations with Hobbes, begins to become more involved with human friends. Hobbes, as Winnicott suggests, is not mourned, but the capacities that were developed during the period of STOs continue to be developed and utilized. Imagine further that Calvin and his parents are able to render the ontological rift inoperative, creating spaces for Calvin to continue to personalize other species.

Before continuing, it is clear, of course, that psychosocial development can include a trajectory wherein the apparatuses of the ontological rift take precedence. We have seen, and continue to see, this as normative and healthy. These theories of development imply that we do not need to animate or personify other living beings or objects to dwell in this world. This seems "true" for many "civilized" Western peoples, given the last 7,000 years of Western civilization. Yet, the climate crisis reveals that this way of dwelling in the world is, in the end, not healthy for human beings or other species. Winnicott (1971) quipped that "the absence of psychoneurosis may be health but it is not life" (p. 100). Western subjects may be deemed to be psychologically healthy, but, if we are captive to the apparatuses of the ontological rift, our subjectivities are not ecologically healthy because of our trajectory toward climate disaster—a disaster that will unhouse human beings and millions of other species. The ontological rift produces dissonant antagonistic relations with other species and the Earth. To dwell ecologically, to experience being alive, depends on living with, as well as learning about and from, other species in terms of their singularities. In other words, experiences of being-in-itself, being-for-itself, and being-for-and-with-others are part of the challenge of surviving and thriving in the precarious habitat that we share with all living beings. Children and adults in Indigenous communities know this, and, perhaps, some people in the West are beginning to see this as well.

Let me return to STOs before addressing the transition to what I call tertiary objects and the subsequent divergence from earlier epistemological principles. As noted earlier, Winnicott (1971) believed that TOs are not mourned because children transition to other activities and more social forms of play. The TO "loses its meaning, and this is because the transitional phenomena become diffused, have become spread out over the whole intermediate territory between 'inner psychic reality' and 'the external world' … over the whole cultural field" (p. 5). This, however, does not mean that transitional phenomena are left behind. For Winnicott, science, art, and religion are adult social forms of transitional phenomena (p. 14). This requires some clarifications. Primary and secondary TOs are under a child's "omnipotent" control,[33] while adult transitional phenomena are shared. This suggests a shared belief in and experience of omnipotence, the evidence of which is found, perhaps, in fantastical, creative thinking in art, religion, and science. I am willing to concede that human beings engage, at times, in omnipotent thinking that has fantastical, pragmatic, and social functions. However, there is an important distinction to be made in the omnipotent thinking associated with PTOs and STOs and that associated with Western Cartesian–Baconian sciences and Western religions and philosophies. As noted above, PTOs and STOs do not include beliefs in having dominion over the object or the inferiority of the object or the superiority of the subject. Early forms of omnipotent (or imaginative) thinking are aimed at securing subjective and later intersubjective experiences of being-in-itself, being-for-itself, and being-for-and-with-others. The latter indicates a recognition of the singularity of the other, like we see in Calvin's perception of Hobbes as a tiger-person. The omnipotent thinking of the Cartesian–Baconian sciences is one of domination of, control over, and instrumental use of other species and the Earth—a denial of the singularities of other species. Also, much of the omnipotent thinking of Western religions may be observed in the anthropocentric, exceptionalist views of some human beings who believe that they have received permanent existential significance from an all-powerful entity while denying the ontological significance of other species and othered human beings. Calvin's STO, by contrast, is an example of imaginative thinking that is personal and anarchic. In Western subjects, this is largely left behind as imaginative omnipotent thinking becomes associated with features of the ontological rift.

There are two other clarifications to be made. First, for Winnicott (1971), "The task of reality-acceptance is never completed" (p. 13). Transitional phenomena provide a resting place from the challenging demands of accepting reality. We might consider that STOs provide not simply a resting place but a space to engage playfully and resonantly with personalized objects. In other words, the space enables children to make use of cultural objects in learning to dwell resonantly with another person-subject. Calvin's engagement with Hobbes entails mutual playing, arguing, making up, and so on, all aimed at dwelling together. The STO, then, is not in this case a resting place but rather a space to work out how to dwell with another person. Let me stress that, in this space, there is no hint of domination nor are there beliefs in inferiority and superiority. By contrast, the transitional phenomena of Western adults represent spaces where collective dwelling is exclusive, whether that means excluding othered human beings and/or excluding other species. Put another way, the ontological rift represents transitional phenomena that are captive to epistemologies dependent on domination and attendant beliefs in the superiority of (some) human beings and the inferiority of other species.

Second, while transitional phenomena may provide a resting place, they also, for Winnicott, serve as a source of primary creativity (1971, p. 11). Here again, we must differentiate between transitional phenomena of childhood and those of adulthood. Primary creativity, which includes aggression, is devoid of beliefs in ruling over the object and the inferiority of the object. In Calvin's creative play with Hobbes, there is no hint of aggression in the form of domination or inferiority. Instead, there is a mutual (creatively imagined) recognition of each other's significance and singularity, which can include moments of aggression (e.g., arguing, fighting). Indeed, primary creativity includes an animistic, personalizing epistemology that is associated with STOs, which means that other objects, like Hobbes, dwell with oneself. This early creativity and epistemology fall under the principles of multinaturalism and perspectivism, which found resonant dwelling. For instance, Calvin's epistemological stance toward Hobbes affirms, in Calvin's imagination, Hobbes's personhood (singular nature), and this affirmation is connected to his motivation to learn about Hobbes as they dwell together in this space. In brief, Calvin's creativity and omnipotence accompany reciprocity, cooperation, and intimate or resonant belonging in the face of shared precarity.

This is not the case for many Western subjects. While there is an incredible amount of creativity in the Cartesian–Baconian sciences, much of it is aimed at dominating the Earth, other species, and othered human beings. Western religions and philosophies are creative and pragmatic and similarly shot through with beliefs in human exceptionalism and dominion over other species who may or may not obtain mere use value. In other words, the creativity of Western subjects, captive to the rift, elides the epistemological principles of multinaturalism and perspectivism in favor of the instrumental use of other species for the sake of human dwelling. Learning is also for the sake of instrumental use. To be sure, many adult transitional phenomena are also reciprocal, cooperative, and intimate, at least with regard to those who share the phenomena. But, when creativity comes to relating to other species, it is one of depersonalization, control, and instrumental use—dissonance. Creativity and aggression are for the sake of a privileged group's dwelling and belonging. For instance, some human beings have exercised their creativity with the aim of the industrial use and slaughter (unhousing) of billions of animals for human consumption. And then many of us uncritically accept this form of creative injustice when we purchase these "products."

Yet another question concerns the term "transition." This term has more than one meaning. It can refer to the movement from one stage or phase of development to another. It can also refer to the transition from a respite from reality toward reality acceptance, which, for Winnicott, is never completed. I believe it is quite plausible that human beings, because of the precarity of life, sometimes struggle to accept reality and need a respite. However, I think the TOs of childhood are part of reality. As noted above, personalizing objects—animate and inanimate—represent an epistemological stance of learning about the object. Yes, there are many illusions, but I am suggesting that infants and children are positioned to learn from, adapt to, and cooperate with objects in their environment. This is not mere fantasy or rest from reality acceptance. Calvin's playing is not simply a respite from the demands of living with other children and adults. He is learning to cooperate with, adapt to, and live with another being qua person, though in his case it is an illusory tiger-person. I add here that the "reality" of the Calvin–Hobbes world is also one of shared gratitude, joy, awe, and so on, which is preparatory for sharing the resonant pleasures, joys, and tribulations of adult living. These experiences with Hobbes are, for Calvin, real, and they are real in the sense of using the social apparatuses of his community.

Yet, as Calvin transitions to leaving Hobbes behind (maturation), a question arises about the transitional phenomena of adulthood. These phenomena are not transitional in the sense of maturation, though some religious transitional phenomena may view life as a transition from an earthly to a heavenly home. Winnicott, I believe, viewed the "transitional" in adulthood primarily in terms of moving from a resting place to reality acceptance, which is, for him, a lifelong individual and collective task. That is, we might say there is a transition between a creative rest from reality to acceptance of reality, which is not an issue of maturation as in the case of PTOs and STOs. There are all kinds of examples where adults engage in shared fantasies (movies, gaming, sports) that may provide a sense of relief from the harsh realities of life. But I think we need to make a distinction between the "transitional phenomena" of Western science, philosophy, and religion and those of Indigenous peoples. Although it is a broad generalization, Western Cartesian–Baconian sciences, religions, and philosophies are not simply areas of shared rest but rather are bent on asserting dominion and superiority over other species and the Earth. There can be shared awe, wonder, and pleasure in the creative work of the sciences, but there is also a constant agitated or unquiet attempt to maintain dominance and superiority over "nature" as against engaging in mutual, reciprocal, and cooperative relations with other species and habitats. Western religions and philosophies, which have supported and legitimated the destructive colonization of innumerable people across the globe, are similarly preoccupied by human exceptionalism and exemptionalism. The transitional phenomena of Western societies, for the most part, produce relations of domination, projecting inferiority and insignificance onto othered peoples, other species, and the Earth. In one sense, the Western transitional phenomena evident in Cartesian–Baconian sciences and Western religions and philosophies represent not a respite from reality acceptance but an ongoing refusal to accept the reality of our existential insignificance and impermanence while projecting both onto other species and othered human beings. Put another way, the apparatuses of the rift can create a space for respite from the harsh realities of life, but they do so at the expense of resonance with other species and the Earth. Our respite fosters dissonant dwelling with other species and the Earth.

Of course, these are large claims, and one can find numerous instances of Western adults who reject domination, superiority, and inferiority. However, any cursory overview of Western history reveals a long and wide trail of destructive attempts to dominate peoples, other species, lands, seas, and so on. Indeed, the climate crisis is a symptom, in my view, of Western transitional phenomena or apparatuses that produce and maintain human privileges, exceptionalism, superiority, and dominion. If there is rest by way of the apparatuses (transitional phenomena) of the ontological rift, it comes at the cost of projecting existential insignificance and impermanence onto other species. But it is a fitful, tragic rest, especially given the present and future realities of the climate crisis.

So, I note that the transitional phenomena of adulthood have nothing to do necessarily with transition to another stage of development or maturity. That said, the idea of maturity and development takes on a particular hue regarding the ontological rift. In the previous chapter, I argued that, while Freud acknowledged the animistic epistemology of children and adults, he nevertheless assigned this to early childhood—something to be surpassed. Winnicott, while more open to illusion in adult living, appears to echo this. The creative play between Calvin and Hobbes is to be left behind for more socially shared forms of play. As I have pointed out above, in the West, for those captive to the rift, play and creativity are shot through with beliefs about domination, superiority, and inferiority. Western psychosocial maturity, in other words, entails splitting off animistic, personalizing epistemologies as well as the principles of multinaturalism and perspectivism. Western adults of the rift may be amused by Calvin and Hobbes, but, when it comes to Indigenous peoples, who operate out of animistic, personalizing epistemologies, Western subjects are often dismissive at best and hostile at worst.

I am now ready to posit the idea of tertiary objects in relation to development and adulthood. Tertiary objects are the narratives, rituals, shared objects, institutions, and so on, of one's social group. These socially constructed objects ideally make possible mutual cooperation aimed at resonant dwelling together—shared being-in-itself, being-for-itself, being-for-ourselves (for and with each other)—in the face of individual and collective shared precarities.[34] Although an oversimplification, there are two tracks regarding the transition to tertiary objects. The first track involves the apparatuses or tertiary objects of the ontological rift. Let's imagine that Calvin, having left Hobbes behind, has internalized the apparatuses of the rift. He engages in all kinds of social activities wherein there is cooperation and reciprocity, though he regards his having played with Hobbes as silly, even childish. Having internalized the apparatuses of the rift, Calvin has a notion of maturity based on reason that eschews the personalization of other species. He now accepts that other animals are not persons, though he treats his family's pet dog with respect and care. Like his parents, Calvin, now a teenager, never considers the issue of care and justice regarding other species. When eating meat his parents purchase and prepare, he never thinks about the animal. Indeed, even to begin to imagine the singularities of the animals slaughtered for his and his parents' well-being would seem ludicrous and sentimental. Calvin, like his parents, is not at all cruel and unfeeling. Rather, having internalized the tertiary objects of the rift, he gives no thought to other species except in instrumental ways. He has unconsciously (encompassing the normative, historical, political, and developmental unconscious) accepted that human dwelling is privileged by way of human superiority over other species, since other species, for those of the rift, are not persons.

The other track is not one of discontinuity with early childhood epistemologies but rather one of integration with tertiary objects. To explain this, I turn to collections of North American Indigenous stories (see Erdoes & Ortiz, 1984; Kerven, 2018). These stories represent an integration of personalizing, animistic epistemologies with sophisticated shared constructions of experience that are necessary for surviving and thriving as a people in their particular habitats—a social ecology of life. Numerous animals and insects are animated or personalized in these stories, which represent beliefs and values about how to dwell in this world. While these stories may have been and are entertaining, that is not their main function. They represent a particular Indigenous group's understanding of the world, its inhabitants, and life's precarities, as well as how to overcome, defeat, or cooperate with other species and forces in the world. Anthropologist Tim Ingold (2022) offers an illustration. A Cree hunter of caribou (an adult Calvin in my imagination)

> is a perceptually skilled agent, who can detect those subtle changes in the environment that reveal movements and presence of animals. … Secondly, he is able to narrate his stories of his hunting journeys. … When the hunter speaks of how the caribou presented itself to him, he does not mean to portray the animal as a self-contained, rational agent whose action in giving itself up served to give outward expression to some inner resolution.
>
> (p. 29)

The caribou is not, to the hunter, a human person, but rather a caribou-person with its own unique sentience (being-in-itself, being-for-itself). I add here that the Cree hunter, while an individual, is inextricably linked to his people, and his hunting is for the sake of their collective well-being.

One may wonder if I am naively suggesting that we become more like Indigenous peoples by giving up our sophisticated technologies and ways of living. Do we sacrifice the epistemological power of science for more ecological ways of living and cooperating with other species? My response is to first approach, briefly, the issue of science. The West has been and continues to be dominated by Cartesian–Baconian scientific methods—apparatuses of the rift. As I have already mentioned, this kind of scientific thinking objectifies and depersonalizes the objects of study. This objectification accompanies beliefs (illusions) of human superiority, anthropocentric privilege, the inferiority of other species, and instrumental epistemology. Most of the time, we think nothing of this type of science, only to be rightly horrified when we learn this approach is directed toward human beings (e.g., Tuskegee experiments on African American males). Part of this horror is related to our tendency to reserve singularity to human persons. Since other species are believed to lack singularity (because they are viewed as mute and dumb), Cartesian–Baconian scientists doing research on animals are free from the constraints demanded by institutional review boards. Although I can hear the groans and snickers from scientists who would view this as simplistic, naive, or silly, the point, nevertheless, is that this kind of science is captive to the rift.

Of course, this approach to science can also be considered in terms of inanimate objects and forces—both visible and invisible. Physicists, for instance, construct complex equations to study matter and antimatter. We would not expect them to personalize or recognize the singularities of the objects they are studying. Cartesian–Baconian science would appear to be fine in this realm, but I wish to problematize this. As long as scientists of the rift split off the impacts of their objectifying knowledge on the environment, this approach cannot be justified as ethical. For instance, there is a great deal of science in fracking and mountaintop removal mining, but this science, which is imbricated with the political-economic apparatuses of capitalism, is also implicated in environmental devastation, impacting human beings and other species. Cartesian–Baconian scientists may not attribute singularities to inanimate objects but, by failing to recognize the singularities of other species and all living creatures' need for a habitable world, they reveal their captivity to the rift. For Indigenous peoples, the land is sacred because they recognize it as the very material foundation for their being-in-themselves, being-for-themselves, and being-for-and-with-others—human beings and other species.

There are scientists who are not captive to the rift and who recognize the singularities of the animals they are studying. Primatologist Jane Goodall (2010) leaps to mind. She spent years studying and protecting the lives of apes. Sy Montgomery's (2016) book on the lives of octopi is another. And such scientists are not just those studying other animal species. Some botanists, as Zoé Schlanger (2024) details, recognize and respect the singularities of plants and trees. These scientists are not integrating scientific methodologies with animistic-personalizing epistemologies noted in Indigenous narratives but they are, for a lack of a better term, making use of epistemologies of singularity. They recognize that the living beings they study possess singularity, and this leads to ethical approaches to learning about their ways of dwelling in the world. These scientists are not captive to the ontological rift. More to the point, they represent a science wherein tertiary shared objects represent an integration of the fundamental epistemological principles of early animistic knowing, namely, multinaturalism and perspectivism—principles that hold the belief in the singularity of the other.

It is not just scientists who demonstrate resistance to the rift by treating other species in terms of their singularities. As mentioned in the previous chapter, Plutarch argued for the rights of animals, and centuries later Porphyry contended that it is our moral duty to extend justice to other animals. In the 19th century, philosophers Jeremy Bentham and Arthur Schopenhauer posited that animals have rights. More recently, philosophers such as Peter Singer (1975, 2023), Andrew Linzey (2009), and Martha Nussbaum (2022) have explored the thorny ethical issues regarding human beings' treatment of other species. There are also strands in Christianity (e.g., Francis of Assisi, Teilhard de Chardin) and Buddhism (Keown, 2000) that express the importance of noninstrumental, even anarchic, caring for other species.

My point is that the movement from secondary transitional phenomena to tertiary objects can proceed in several directions. The first entails internalizing the apparatuses of the rift, which leads to a splitting-off from animistic, personalizing epistemologies that are present in PTOs and STOs, though with significant differences. A second possibility is the integration of animistic epistemologies with more sophisticated epistemologies of tertiary objects, which is evident in Indigenous narratives and rituals. A third possibility involves retaining the principles of multinaturalism and perspectivism in tertiary objects, which means rejecting Cartesian–Baconian depersonalizing approaches to science and acknowledging the singularities of other species, as well as the necessity of a habitable biodiverse environment for collective survival and thriving.

Conclusion

The perspective offered in this chapter has several important implications. First, one of the consequences of the ontological rift has been the exclusion of other species from political philosophies. This was exacerbated by the scientific revolution in which "nature" and species other than humans were reified and depersonalized with the aims of control, mastery, and exploitation. This trajectory and its attendant apparatuses have implications for psychosocial development. Erik Erikson wrote that all parents "aim" their children. I would add that all societies aim their members. Thus, Western subjects, as they develop psychosocially, internalize the beliefs and values of the ontological rift, which fosters antagonist and controlling forms of relating to othered human beings, other species, and the Earth ("nature"). To offer a psychosocial developmental perspective that renders the rift inoperative by way of anarchic care is a political act in the sense that other species and the Earth are necessarily integral to resonant relations, political dwelling, and psychosocial development. Imagine a real Calvin who sees and constructs Hobbes as a tiger-person and, therefore, a citizen of the polis. As Calvin grows up, we can imagine him rendering the apparatuses of the rift inoperative, thus creating a space for other species qua persons to be of and belong to the polis.

A second and related implication concerns the issue of justice. The ontological rift confines the issue of justice to those on the side of superiority, which means othered human beings, other species, and the Earth are excluded from these discourses. The psychosocial developmental view offered in this chapter expands the discourse to other species and the Earth. To return to our fictional Calvin and Hobbes, each sought to give each other what was due, and, when they argued, they aimed at repair—all in Calvin's imagination, of course. This reciprocal, cooperative, resonant play is an early expression of working out just relations, which includes repair. If we imagine Calvin's developmental trajectory in terms of tertiary objects that render inoperative the apparatuses of the ontological rift, then just relations extend to other species and the Earth.

A third implication concerns psychoanalysis. Analysts of varied persuasions, over the decades, have argued that psychoanalysis needs to include the political in its theorizing and in therapy, when appropriate. Less attention has been paid to the inclusion of other species. The climate crisis reveals that all living beings are mutually dependent on this one biodiverse Earth. That is, we are dependent on other species for dwelling, for survival and flourishing. And it appears that other species are dependent on human beings to take care of the environment. The implication is that psychoanalysis must be an ecological science of the human being situated in and dependent on a biodiverse Earth. We, like all living beings, are semiotic creatures who face the precarities of life and the reality of existential insignificance. We dwell and belong with other living beings. To focus simply on the embodied psyche divorced from other species inadvertently replicates the ontological rift and fosters dissonance. To bridge this rift, psychoanalysis must render inoperative the apparatuses of the rift by choosing an ecological science of human psychosocial development wherein personalizing epistemologies remain integrated with other forms of knowing.

A fourth and likely controversial implication concerns science. The Cartesian–Baconian sciences operate out of instrumental, reifying epistemologies that are dependent on apparatuses of the ontological rift. Is it possible to reimagine science so that it embraces the singularities of other living beings?

Can there be animistic sciences that accept and operate out of the belief in the singular aliveness of other beings? Can there be a science that embraces anarchic care with regard to its subjects? It is clear to me that psychoanalysis, which many view as a human science, can embrace the notion of anarchic care for patients. But can it embrace anarchic care and the singularities and aliveness of other species? Perhaps more disturbingly, can there be an animistic psychoanalytic theory that shapes practice? This chapter is a step in this direction.

Let me end with a brief thought. The poet W. H. Auden, in giving us a stark choice that is even truer today, writes, "We must love one another or die" (Messerly, 2014). In a different vein and context, Valerie Kaur (2020) writes, "This is our defiance—to practice love even in hopelessness" (p. 241). A premise of this chapter is that anarchic love, in rendering the ontological rift inoperative, creates spaces for children to appear in their singularities and to actualize their potentialities, especially regarding resonant dwelling with other human beings, other species, and the Earth. Without this anarchic care, children either perish or develop selves that survive but do not thrive. This anarchic love is founded on the principles of multinaturalism and perspectivism, as well as the personalizing epistemologies that are methods of learning about and learning how to dwell with the singularities of others. A psychoanalytic ecological theory of human development extends this anarchic care to other species and the Earth. This extension is our defiance—a defiance that renders the apparatuses of the rift inoperative.

Notes

1 For Agamben, "inoperativity" means the deactivation of the functioning of the apparatuses, which does not mean that these apparatuses do not impact or shape the individual's unfolding (Prozorov, 2014, pp. 31–34). Martin Luther King Jr. rendered the apparatuses of white supremacy inoperative, though these apparatuses continued to function. King was attacked and eventually killed. So, to render the apparatuses of the ontological rift inoperative means that, while the way we dwell in the world changes, the rift will continue to have its negative effects on othered human beings, other species, and the Earth.

2 As mentioned in the previous chapter, Western philosophies and theologies have, by and large, excluded other species from political consideration as necessary members of the polis. In this definition, "other species" means other species of flora and fauna.

3 Horkheimer commented that "the future of humanity depends on the existence today of the critical attitude" (as cited in Wolin, 2016, p. 231). This is crucial as cultural critics, such as Giroux (2012a, 2012b) and Dufour (2008), note that, with the rise of neoliberal capitalism, there has been an increase in the formation of subjects or citizens who are acritical of the systems that contribute to their suffering and the suffering of others.

4 Philosopher David Livingstone Smith (2021) explores the notion of dehumanization and its destructive impacts on human beings. While I find his work helpful and important, I prefer the term "depersonalization" because it can refer to our relations to and behaviors toward other species. In other words, depersonalization entails denying others, whether human beings or other species, personhood. Dehumanization applies solely to human beings and their relations with each other.

5 Philosopher Byung-Chul Han (2019) writes that "all finite beings are surrounded by otherness. … Power is the capacity of the living not to lose itself while being entangled in the other in manifold ways" (p. 50). More positively, power (or agency) is the living creature's capacity to assert and affirm being-in-itself and being-for-itself.

6 The term "narcissism" continues to be used in psychoanalysis and other fields. The concept is derived from a Greek myth in which Narcissus falls in love with his image. Generally speaking, the notion is used in reference to an early period of psychosocial development, as well as a kind of pathology. There are inevitable questions regarding whether infants or children fall in "love" with their image. I prefer the terms being-in-itself and being-for-itself instead of narcissism because they bridge the ontological divide. There is no evidence that other species go through a stage of narcissism, though many believe all humans do. We can safely posit that all species in their singularities have the "drive" (for lack of a better word) to being-in-itself and being-for-itself, which is aimed at survival and flourishing. It would, in my view, be better to say that narcissism, as a pathology, represents distorted ways of being-in-itself and being-for-itself in relation to dwelling with others.

7 Byung-Chul Han (2019) ascribes power to a living creature's ability to affirm itself, but for human beings this is complicated by the fact that human beings are social and communal animals. Positive power is the exercise of parents' caring personal recognition and treatment of a child that affirms the child's agency in being-in-itself and being-for-itself.

8 Lyndsey Stonebridge (2024, p. 128), in writing about the life and work of Hannah Arendt, notes that Arendt uses the phrase *volo ut sis* (I want you to be) in discussing love, moral reasoning, and politics. Parents' anarchic care can be understood as *volumus ut sis* (we want you to be), which creates a space for infants to actualize being-in-itself and being-for-itself.

9 I wish to stress that Agamben's differentiating between human animals and other species does not imply a hierarchy or belief in the superiority of human beings. Coming from a different but related angle, philosopher John Gray (2013) wrote, "The distance between human and animal silence is a consequence of the use of language. … But if animals lack this interior dialogue, it is not clear why this should put humans on a higher plane" (pp. 163–164). I add here that it is not clear to me that other animals do not have some capacity for impotentiality. I am confident I have observed numerous times our cat, Grace, "preferring not."

10 What distinguishes or differentiates humans from other-than-human species (or from each other) does not imply superiority, ontological or existential significance, privilege, etc.

11 The notion of resonance has been used by psychoanalysts most frequently in the context of therapeutic communications (see Coburn, 2001; Levine, 2009a, 2009b; Stone, 2006; Välimäki, 2015), which Rosa (2019, p. 167) recognizes. The notion of resonance, while not directly used, is also evident in the philosophical works of Alfred North Whitehead (Faber, 2023; harmonic vibrations), Martin Buber (1958; I–thou relations), Teilhard de Chardin (1978; vitalization of matter and spirit of the Earth), and Maurice Merleau-Ponty (1964; syncretic sociability). Rosa's in-depth work on resonance is particularly useful for this and subsequent chapters.

12 Rosa (2020) is not restricting the notion of resonance to human beings alone (p. 35) or to living beings, though his focus is understandably on human beings. One of the features of resonance, for Rosa, is adaptive transformation. For any living being to survive and thrive, it must adapt to the environment upon which it is dependent. It must resonate with the environment, which will include some of its inhabitants.

13 For Rosa (2019, p. 169), it is difficult to be related resonantly to the world if one is in an antagonistic relation to an abstraction such as "nature." Rosa identifies this as alienation, which I also view as dissonant relations and subjects.

14 Rosa briefly and tentatively mentions in passing, "A baby, perhaps even as an embryo, experiences and lives in resonances, long before it can say an 'I'" (2019, p. 449). I develop this a bit further, arguing that researchers have provided evidence of pre-birth organizations of experience and that these organizations can be understood in terms of the notion of resonance.

15 My focus is on human psychosocial development. However, I would contend that resonance is a concept that is applicable to all living beings. In other words, resonance is a feature of semiosis and singularity. The work of Robert Penrose, Stuart Hameroff, and Subhash Kak (2017) on the relation between sentience and the material world is also relevant here. Briefly, they argue that material existence manifests resonance at the quantum level, which indicates that resonance is the basis of animate, sentient semiotic activity. All living beings are semiotic. While resonance applies to material existence, there is a distinction between resonance at the quantum level and resonance in relation to semiosis or life. I suggest that the distinction is that living creatures can experience resonance, whereas inanimate objects cannot. Put differently, resonance from the quantum level (atoms and subatomic particles) to molecules, mitochondria, cells, bacteria, simple organisms, animals, etc., represents increasingly complex (not higher) organizations of resonance. The implication here is that resonance is a foundational feature of material existence and all living beings. As Rosa (2019) writes, "The primal form of existence is a relationship not of alienation but of resonance" (p. 258). This does not mean there is no dissonance; dissonance exists but is an epiphenomenon.

16 Christopher Bollas (1987) coined the term "transformational object," which he associates with early infancy. Bollas writes that he wants "to identify the infant's first subjective experience of the object as a transformational object" (p. 14). Of course, the "object" is the parent who "alters the infant's environment to meet [the infant's] needs" (p. 15). The parent object is not an object in the sense of the infant being able to differentiate between this or that object. More precisely, the parent object is "a process that alters the infant's experience" or subjectivity (p. 13). In other words, the "object is 'known' not so much by putting it into an object-representation, but as a recurrent *experience of being*" (p. 13, emphasis mine). This early experience of being, Bollas contends, is the basis for later experiences of being in adulthood, such as when one is moved by music or a sunset. While Bollas's view is different from the notion of resonance, there is a similarity in that both indicate a relation of being and belonging. That said, I prefer resonance because it is more interactive than passive. Bollas's transformational object is in relation to a passive infant (or adult), except perhaps to the degree that the infant surrenders to the object. The term "resonance" has more of a sense of mutuality.

17 Philosopher Merleau-Ponty (1964) argued that early infancy entails the development of a corporeal schema in relation to the other. He called the first phase pre-communication, by which he meant prelinguistic. This phase "is not one individual over against the another but rather an anonymous collectivity" (p. 119). While Merleau-Ponty does not address pre-birth, his recognition of the early perceptual-relational field of infants is fitting here because it closely parallels the notion of embodied, relational resonance.

18 Rosa (2019) appreciates yet disagrees with Alex Honneth's (1995) theory of recognition by arguing that recognition and resonance are not equivalent. Indeed, he contends that his theory of resonance "surpasses recognition theory in its explanatory potential" (p. 196). The stress is not simply on recognition but on a type of

"dynamic event" that manifests "a vibrant responsive relationship" (p. 196). Rosa is not denying the importance of recognition; however, his focus is on the relational experience of resonance. I suggest that the idea of parents' personal recognition that is embedded in anarchic caring attunements to infants' assertions makes resonance possible. That is, the theories of recognition and resonance come together in these caring interactions.

19 Otto Rank (1929/2014) argued that "*analysis turns out to be a belated accomplishment of the incompleted mastery of the birth trauma*" (p. 5, emphasis in original). Rank, I believe, is indicating that the nascent, pre-birth ego is unable to "master" the experience of this new reality, hence his use of the term "trauma." I am not convinced that this first experience of being unhoused is traumatic. Yes, birth is, as William James noted, confusing, disruptive, painful, etc., but it is not necessarily traumatic. If we follow Rank, then every human being needs therapy unless they have found other means to "master" the birth trauma. It is more likely that the preferences (pre-birth semiotic organizations) for the warmth of the mother's body and for swaddling soothe infants' entry into this new world.

20 In Chapter 4, footnote 9, when addressing eco-transference, I identify some problems with positing omnipotent thinking in early childhood. For now, I am retaining the idea of omnipotent thinking, but, as I explain in Chapter 4, the notion of omnipotent thinking in early childhood is problematic when we consider the development of primates. It is also troubling because it is better associated with the apparatuses of the ontological rift, which are present in but not yet internalized by infants. While the use of the term is widely used in psychoanalysis, my preference is to replace it with the broader term "imagination."

21 The notions of self-esteem, self-respect, and self-confidence are used by Axel Honneth (1995) in his political philosophy. He does not locate these in early psychosocial development, but I do here because the later actualization of political agency depends, in my view, on these earlier semiotic organizations of self-respect, self-esteem, and self-confidence. I add here that these senses of self can also be seen in terms of the drives of being-in-itself and being-for-itself, which will later be connected to being-with-and-for-ourselves.

22 Rosa (2019) also points out that there is "overwhelming evidence that children are healthier, more resilient, and more creative, and have higher expectations of self-efficacy, when they are in regular contact with animals and expansive natural spaces in which they experience sensual/practical resonant relationships, the therapeutic value of which for adults has also been well documented" (p. 272). In my view, this illustrates the benefits of rendering the ontological rift inoperative, which I will return to in Chapters 5 and 6.

23 Winnicott did not differentiate between the TOs of childhood and those of adulthood, which raises all kinds of questions. The developmental achievements and complex psychological and relational realities that take place between infancy and adulthood are huge. This in itself would demand differentiation between types of objects. In this case, I consider the first TO the PTO—an object that is associated with presymbolic modes of organizing experience. For critiques of Winnicott's theory of development, see Applegate (1989), Brody (1980), Flew (1978), Greenberg and Mitchell (1983), and Litt (1986).

24 Pruyser (1983) suggests that the TO is shared to the extent that the object is often one that is "found" in the cultural realm of family life. Moreover, parents and siblings tolerate and accept the child's use of the object. While I agree with this view, it is clear that Winnicott does not believe that the earliest Tos, or what I call PTOs, are shared or intersubjectively held, at least not during early infancy.

25 Winnicott did not assign care simply to the mother. A mothering environment could be accomplished by fathers (and others) as well.
26 One might wonder what happens to primary (and secondary) TOs as the child grows. In the course of human development Winnicott (1971) argued that the TO "is not forgotten and it is not mourned. It loses its meaning, and this is because transitional phenomena have become diffused: they have become spread out over the whole intermediate territory between 'inner psychic reality' and 'the external world as perceived by two persons in common,' that is to say, over the whole cultural field" (p. 5).
27 As noted in Chapter 1, since Freud the psychoanalytic tradition has tended to associate animate thinking with early childhood and so-called primitive people or savages. It is a type of thinking and relating that is to be overcome, surpassed by reason. This results in overlooking the evolutionary and adaptive function of animism. Moreover, so-called primitive people who "animate" other beings and objects seem to have dwelled quite well and sustainably in their habitats for thousands of years.
28 Hannah Arendt (1958), in her political philosophy, uses the terms "space of appearances" and "speaking and acting together." In my view, the STO represents a pre-political space of appearances. This space is free of the ontological rift.
29 I do not believe we have to posit omnipotent thinking in children's play. It is in my view simply imaginative animistic interactions.
30 Viveiros de Castro (2017) argues that Western epistemologies tend to desubjectify the other. "The form of the Other," he writes, "is the thing. Amerindian shamanism is guided by the inverse ideal: to know is to 'personify'" (p. 60). To personify is a method of coming to know the other, while instrumental and objectifying epistemologies are about knowing the other in terms of its use value.
31 In explicating perspectivism and multinaturalism, anthropologist Viveiros de Castro (2017) contends that "virtually all peoples of the New World share a conception of the world as composed of a multiplicity of points of view. Every existent is a center of intentionality which apprehends other existents according to their respective characteristics and powers" (p. 55). He states further that "multiplicity can be taken as a kind of plurality [and] Amazonian multinaturalism affirms not so much a variety of natures as the naturalness of variation—variation as nature" (p. 74).
32 There are, not surprisingly, a great many illusions within scientific thought. The most dangerous illusion of the Cartesian–Baconian sciences is the belief that human beings can control or master nature. Indeed, the very idea of "nature" as distinct from human beings is an illusion.
33 Winnicott (1971) contends that "some abrogation of omnipotence is a feature from the start" (p. 5). So, whether we are addressing PTOs, STOs, or adult transitional phenomena, the belief in omnipotence is partial, in part because even infants have to adapt to reality, and this adaptation requires the abrogation of omnipotence. Having said this, I have argued that we do not have to agree with this view. We can simply say there are imaginative semiotic capacities in early life.
34 Tertiary objects may function as transitional phenomena in adulthood, but these objects are not identical to primary or secondary TOs. They are not solipsistic or under the omnipotent control of an individual. Moreover, they have many other social functions, such as mutual cooperation and reciprocity toward pragmatic social goals.

Chapter 3

Trauma, Human Beings, and the Silence of Othered Species

John was in his seventh decade. We met because he wanted to talk about his son. John was concerned and wondered how he might offer help without repeating his past mistakes. I do not remember how we arrived at this conversation, but John said he had been a hunter when he was a child and young adult. He loved going with his dad on trips to hunt deer, elk, and other animals. This love overrode his sense of being troubled by the suffering of the animals they killed. "When I was in my mid-20s," John said, "it suddenly came to me that I didn't love hunting at all. I just liked spending time with my dad in the woods." From that day forward, John never hunted again, though he spent time with his children outdoors. "I just hated to see these animals suffer. It pained me and I no longer wanted to be involved in a sport that led to unnecessary suffering." Because John was there for two consultation sessions, and perhaps because I was reluctant to know the answer, I did not ask at the time if this ethical stance had led him to stop eating meat altogether, given that almost all meat is derived from the horrific suffering arising from industrial farms and slaughterhouses.

John's sensitivity or caring imagination reminds me of Jack London's (1903/1981) story *Call of the Wild*. This story involves the interaction between human beings and dogs, wherein dogs were horribly abused and then later recovered through the tender (anarchic) care of two men. Jack London and John recognized that other-than-human animals not only suffer but can also be traumatized, something that many pet caregivers know well. At best, we tend to attribute suffering and trauma to animals we identify with, while often ignoring animals and other creatures such as rodents, reptiles, and insects. One might be surprised by the latter category. Does the notion of trauma apply to ants and other insects? Do plants and trees "experience" trauma? Do only sentient, more complex beings such as dolphins, horses, pigs, dogs, and cats experience trauma, and, if so, how would we seek to understand and respond to these traumas? These questions move us into the complex and contested category of sentience and consciousness, which it is neither possible nor necessary to address here, because we are, despite emerging sciences regarding other species and their semiotic organizations of dwelling in the world, largely ignorant of other living beings (see Birch, 2024).

DOI: 10.4324/9781003518013-4

A psychoanalytic reader may wonder why a concept forged in relation to the consulting room and embedded in human psychological theories pertains to other species. Are not psychoanalysts and other therapists focused on understanding human mental suffering, its sources, and the therapeutic processes to address psychological traumas? To what end(s) is it necessary or helpful to ascribe the notion of trauma to other species? Of course, the preponderance of the literature in the psychological sciences addresses trauma in terms of human beings. But the core motivations, in my view, of psychoanalytic explorations, beginning with Freud, are to understand and seek to alleviate psychological suffering, which, as Freud acknowledged, includes other-than-human animals.[1] In terms of human beings, the talking cure, getting the symptom to speak, emerged out of a concern for persons who are suffering and, as a result, are captive to their symptoms. But the question remains: why should psychoanalysts and others in the psychological sciences consider exploring the issue of trauma as it pertains to other species? Let me offer several responses. Recall that in Chapter 1 I noted that Freud believed human beings to be animals and, in part, believed he "was diminishing the gulf" between humankind and other species (1939/1964a, p. 100). Elsewhere in his work, Freud (1900/1953, pp. 131–132) noted that animals have dreams and are, to some degree, capable of consciousness (1915/1957b, pp. 179, 189) and, therefore, clearly suffer (1918/1955b, pp. 16, 26). Following Jessica Benjamin (1995), we could say that Freud's desire to diminish the gulf represents a dialectical cognition of likeness in difference and difference in likeness. We are both different from and like other living beings in terms of sentience and suffering. This dialectical tension with regard to recognizing our likeness to other living beings suggests we consider the idea that other living beings can experience trauma. An impatient reader might agree that we ought to recognize the suffering of other species and that we are animals as well but ask why it is necessary to use the concept of trauma when considering other species.

At the risk of further frustration, let me turn to another innovation of psychoanalysis, initiated by Bertha Pappenheim's plea to Freud to listen to her (Breger, 2000, p. 105). Freud developed a method of listening to the unspoken in the symptoms of suffering. If analytic listening can entail inviting the symptom to speak, then, ideally, suffering may be alleviated and the person's agency and freedom affirmed. Frantz Fanon and others furthered this by listening to the voices of the oppressed with the aims of (a) "to '*consciousnessize*' [the patient's] unconscious, to no longer be tempted by a hallucinatory lactification" and (b) "to enable [the patient] to choose an action with respect to the *real source of the conflict*, i.e., the social structure" (Fanon, 1952/2008, p. 80, emphasis mine). When it comes to other species, there is a great deal of silence and a failure to listen to their varied sufferings. It is a question not simply of whether the subaltern can speak (Spivak, 2004) but of whether we can discover ways to listen to and understand the communications of other species. I am not suggesting that psychoanalysis is a method applicable to caring for other species. However, an animistic psychoanalysis can adopt an attitude of listening to and attempting to understand the traumas of other species. If we, as indicated in Chapter 2, recognize the singularities of other species, then it seems to me that we ought to recognize and care about their suffering and well-being. This is especially the case given the current climate polycrisis. Personally, I believe humans have an ethical or categorical demand to care about the suffering of other species, regardless of whether there is a climate crisis. But, even if we reject that demand, it is instrumentally wise or pragmatic to care about the suffering and well-being of other species because our existence depends on a biodiverse Earth. Just as Fanon used psychoanalysis to listen to the silenced voices of the oppressed, and just as his activism led him outside the consulting room, so too the climate crisis invites us to extend our concern for the silence of other species, listening with a third ear to their cries.

Add to this the value of psychoanalysis in understanding why human beings fail to listen, whether that is the countertransference exhibited in the consulting room or exploring how human beings overlook the voices and experiences of marginalized peoples (Gherovici & Christian, 2019; Layton et al., 2006). In addition, analysts can use psychoanalytic theories and conceptual tools to explain why so many of us defend against seeing and heeding the signs and sufferings—human or otherwise—related to the climate crisis (Hoggett, 2012, 2019; Weintrobe, 2013, 2021). And, as indicated in Chapter 1, we can use psychoanalytic concepts to help us understand why we do not attend to the utterances and experiences of other species, which is related to the normative, historical, and political unconscious.

Fair enough, but do we need to change our views of trauma or psychoanalytic theory to recognize, understand, and care about the suffering of other species? If we wish, like Freud, to close the gap between other species and human animals, then we need to have concepts that include, to some degree, other species who are like us yet different. Put another way, a psychoanalytic anthropology should not only bridge the consulting room and the social-political realm; it also should attempt to span the ontological rift between human beings and other species. If we listen to the sufferings of other species, perhaps our views on trauma will change.

There is one other question to raise, which I hinted at above. If all life is semiotic, does that mean that all living beings are susceptible to or experience trauma? Some botanists observe the distress of plants and trees (see Schlanger, 2024), though the question remains regarding whether they "experience" or are conscious of pain or trauma. I recall listening to a philosopher, perhaps Kantian, say that ethics does not apply to insects since they (like plants and trees) lack neural connections and, thus, do not experience pain (or, presumably, trauma). But how does he know? Any person can observe the distress of insects that are dying. As noted above, the suffering of those animals we deem to be sentient and, better, of animals we are fond of is more obvious to us. When I get to the definition of trauma, it will be broad enough to include all living beings. I believe it is better to err on this side because, in doing so, we do not foreclose discourses and spaces of ethics and justice since all living beings suffer.[2] Yes, it may complicate our lives considerably to do so, but in human history there have been too many times when othered human beings and other species have been (and are) excluded from the zones of care, ethics, and justice by way of the ontological rift. More to the point, extending the notion of trauma to all living beings means creating a stance of listening to and doing the challenging work of understanding their particular and varied sufferings, as well as the sources of and remedies for their suffering (Birch, 2024).

To make my case, I begin with a discussion regarding the notion of trauma and its varied characteristics as it pertains to human beings. In brief, human trauma, I argue, represents the eclipse of personal recognition and care, resulting in (1) a profound sense of being unhoused, (2) an undermining of one's resonant senses of being-in-itself, being-for-oneself, and being-for-and-with-others (dwelling together), and (3) an eclipse of one's associated capacities for agency, self-efficacy, and freedom. From here, I amend the notion of trauma with the aim of including other species, recognizing that the specifics of trauma are unique to each species. While there are differences in the specifics of trauma among other species, I argue that an inclusive view of trauma entails a semiotic crisis of a living being's assumptive world, which entails a sense of being unhoused and an eclipse of resonance sense of being-in-itself and being-for-itself. For instance, the traumas of precocial asocial species, such as reptiles, and altricial social species, such as human beings and dogs, differ considerably. In this section, I illustrate this emended version of trauma, relying on Jack London's story about Buck (a domestic dog), whose world is shattered after he is brutally abducted and shipped to the Alaska Territory. This will include identifying the varied sources of Buck's traumas, as well as the source of his recovery. The chapter ends with some implications of this perspective for psychoanalysis.

A few words of caution are necessary before beginning. The history of trauma is complex, spanning other disciplines. My aim is to offer a general view of the idea of "trauma" and its characteristics. Second and relatedly, I intend to frame trauma in light of the work detailed in Chapter 2 with an emphasis on semiotic experiences of resonant dwelling (being-in-itself, being-for-itself, and being-for-and-with-others). Third, some psychoanalysts (Hoggett, 2012, 2013, 2019; Lertzman, 2017; Orange, 2017; Weintrobe, 2013) have noted that the climate crisis has and will result in varied traumas to human beings. In her acclaimed book on a radical ethics for the climate crisis, Donna Orange (2017) recognizes that "the climate crisis affects us humans unequally" (p. 18). Some groups of human beings, as a result of differences in class, race, sex, and political power, will bear the brunt of climate-induced trauma. Orange's radical ethics does not appear to address the ontological rift between human beings and other species. She does, though, make occasional references to other species. Sally Weintrobe (2013) does recognize the rift, yet she seems to tacitly include other species that may also possess a capacity for empathy (p. 207), suggesting that they can experience suffering and trauma. But what about other species lower on the taxonomic ladder? Do we confine radical ethics only to human beings? Do we only consider the suffering of those animals whom we observe to be or we claim to be sentient? I believe a radical ethics vis-à-vis psychoanalysis and the climate polycrisis is necessarily inclusive of all living beings since our dwelling in the world depends on a biodiverse Earth. This means, as indicated above, that a radical ethics for psychoanalysis includes understanding and responding to the traumas of human beings and other-than-human beings.

Trauma: An Overview

Leading figures in researching trauma, Bessel van der Kolk, Lars Weisaeth, and Onno van der Hart (1996), note, "People have known that exposure to overwhelming terror can lead to troubling memories, arousal, and avoidance" in human beings (p. 47). These authors also point out that psychiatry in the 19th and 20th centuries "has had a troubled relationship with the idea that reality can profoundly and permanently alter people's psychology and biology" (p. 47). Since the days of Freud and earlier, there have been varied ideas and arguments about the sources of trauma and madness—an extreme form of and response to psychological suffering (Gilman, 1985a, 1985b; Porter, 1991a, 1991b; Scull, 2015). Indeed, in Freud's works, there appear to be two differing sources of trauma. Jean-Martin Charcot sought to examine and list the symptoms of "hysteria," while Freud (1897/1962) argued that the source of these symptoms was childhood experiences of psychological trauma, a view he later set aside in favor of fantasy (Herman, 1992, pp. 10–14; see also Fassin & Rechtman, 2009, pp. 30–39). In Freud's (1925/1959a) own words,

> With my female patients the part of the seducer was almost always assigned to their father. I believed these stories, and consequently supposed that I had discovered the roots of subsequent neurosis in these *experiences* of sexual seduction in childhood. … I was at last obliged to recognize that these scenes of seduction had never taken place, and that they were only *fantasies* which my patients made up.
>
> (p. 34, emphasis mine)[3]

Otto Rank (1929/2014), initially a disciple of Freud, posited that birth itself was a source of trauma, and, therefore, trauma was not simply the result of childhood experiences of impingement, deprivation, or fantasy. An existential implication of this view is that to be human means beginning life with trauma, which is clearly arguable.[4]

As differing ideas emerged, the search for the sources of trauma included questions about how individuals psychologically cope with trauma. Pierre Janet's notion of dissociation and Freud's notion of repression are used to depict how unbearable experiences are psychically managed (Gay, 1988). World War I sparked greater interest in understanding traumatic experiences and developing various remedies. The psychological sufferings of soldiers (shell shock) during and after World War I, World War II, the Vietnam War, and other recent American wars furthered the clinical research on traumatic experiences—post-traumatic stress and moral injury (Gray, 1970; Hillman, 2003). Recent research on trauma includes not simply how trauma impacts individuals' memories, perceptions, defenses, character, and so on (see van der Kolk, 2015) but also its impact on society and other generations (see Epstein, 1988; McFarlane & van der Kolk, 1996; Wolynn, 2017). While it is clear that in all cultures and societies there are human beings who have experienced trauma (deVries, 1996), how that trauma is understood and responded to varies across time and space.

This said, interestingly, in Western cultures during the last 50 years, as Didier Fassin and Richard Rechtman (2009) note, the notion of trauma became unmoored from its clinical location and began to be used in the larger society to refer to any painful, disturbing, or triggering experiences, which are not necessarily unbearable.[5] The concept, then, became pervasive in social discourses and has had political implications (pp. 27–30). While the social-political aspects of the common use of the notion of trauma are important, that is not my focus here. Instead, I provide some general comments about the attributes of trauma as it refers to human beings before using the term in light of the suffering of other-than-human species.

Decades ago, Hanna Segal (1957) understood trauma to comprise the loss of symbolic function, which means that an unbearable experience associated with an overwhelming event is organized semiotically, not semantically. For instance, a child who is traumatized is not able to organize the experience through symbolization, though the experience is organized by way of presymbolic semiosis (LaMothe, 1999). In other words, the traumatized person is psychologically overwhelmed, making it difficult to organize the experience autobiographically. When this occurs, the trauma remains outside of symbolic meaning and, in many cases, biographical memory (van der Kolk, Weisaeth, et al. 1996; Williams & Banyard, 1999). Today we use the terms "post-traumatic stress" and dissociation to refer, in part, to experiences that remain largely outside autobiographical or semantic memory systems (Bremer & Marmar, 1998; van der Kolk, van der Hart, et al., 1996; van der Kolk, Weisaeth, et al., 1996), though they remain part of eidetic memory (e.g., flashbacks).

Using different language than Segal, Cathy Caruth (1996, 2014) claimed that trauma is identified as horrific events or crises in a person's life that are unassimilated in a person's consciousness and narrative. Caruth (1996) argued that these are "unclaimed experiences" (p. 4). Unclaimed experience is another way to conceptualize semiotically versus semantically organized experiences. But it is not simply how trauma is organized that is important. Caruth added that the core of trauma is "the lack of support, of help, of comfort; being utterly left alone with the experience and having no one to listen" (p. 202). The absence of support and help is not simply a part of the immediate experience of trauma; it is also experienced after the trauma—exacerbating the person's sense of aloneness. Susan Grand (1997, 2000) held an analogous view, calling trauma an experience of catastrophic loneliness that can result from a horrific experience or cumulative deprivation (e.g., absence of personal recognition, emotional nurturance and support, food, and other psychosocial needs; Khan, 1963).

Connected to this catastrophic loneliness or alienation is the concomitant shattering of one's assumptive world by way of betrayals (Janoff-Bulman, 1992) and a sense of the irreparability of relationships (Bromberg, 2001, p. 902). Like Janoff-Bulman, Fred Alford (2013) argued that "trauma takes away our confidence in the existence of a stable, ordered, and meaningful existence" (p. 10). The shattering of one's assumptive world of trust, fidelity, stability, and security and the corresponding belief in the irreparability of relationships are attended by a profound and global sense of distrust and hopelessness (Freyd, 1996), which undermines a person's capacity for present and future intimate relationships.

This can be further understood by turning to psychosocial development. Developmentally, the very capacities for and use of language and symbols to organize experience depend on caring relations of trust and fidelity created by the personalizing attunements of good-enough parents. Relations of trust and fidelity enable infants to actualize their potentialities, including their securing a resonant, embodied-relational sense of being-in-themselves, being-for-themselves, and being-with-and-for-others. This environmental or global semiotically organized trust and the expectation of fidelity are their assumptive world, and all of this accompanies children's ability to be vulnerable for the sake of receiving affection, love, and so on. That is, to receive and experience care, a child's nascent agency must exercise a willingness to be open or vulnerable, which requires trust in relation to a loyal caregiver. Attachment theorists (Cozolino, 2006; Fonagy, 2001; Fonagy & Target, 1997; Gergely & Unoka, 2008; Holmes, 1996) have long observed that trauma is a form of interpersonal betrayal that disrupts children's capacities for relational trust; this, in turn, undermines their capacities for self-reflexivity and organizing coherent narratives.

In short, trauma shatters a child's assumptive world of trust and fidelity in relation to their caregivers, leaving experience unclaimed and the very inkling of vulnerability as exceedingly dangerous. Put differently, trauma occurs at the point of extreme vulnerability. When this occurs, a child's vulnerability becomes linked to distrust, intolerable anxiety, and catastrophic, existential loneliness. The experience is unclaimed because vulnerability is required to do the work of linguistically processing the trauma with another trusted person, but any movement toward vulnerability heightens a traumatized child's or adult's anxiety and distrust. Invulnerability or, better, the avoidance of vulnerability becomes an unconscious strategy for protecting against further trauma, yet, tragically, the result is that the experience remains unclaimed and the person remains alone and lonely—bereft of care and trust.

Let me complicate this further. It is not that semiotic and symbolic capacities in organizing experience are two radically separate systems within one's assumptive world. Long before symbolization is taking place, infants are organizing experience semiotically. They have a semiotic, presymbolic assumptive world of resonant trust. Relying on the work of Merleau-Ponty and Fanon, Athena Colman (2022) argued that infants first develop a corporeal schema that "can be understood as having to do with a pre-representational [semiotic] mode of experience through which we continually, but without conscious effort, dynamically orient and restructure our body in the world according to our projects" (p. 128). Similarly, philosopher Mark Johnson (1987; see also Lakoff & Johnson, 1999) claimed that embodied constitutional structures exist "preconceptually and nonpropositionally in our ongoing meaningful organization of our experience, understanding, and reasoning" (p. 40). Semiotically organized embodied schemas become intertwined with more complex symbolic organizations of our earlier assumptive worlds, and *both are contingent on consistent relations of care and trust.* Trauma undermines embodied pre-representational and symbolic modes of organizing experience. In other words, trauma is disorienting and painful precisely because it is in radical opposition to the very foundations that make organizing experience possible—namely, caring interpersonal relations—leaving victims struggling to find ways to survive alone.

Shifting to the discussion from Chapter 2, trauma as disrupting embodied semiotic and symbolic organizations of experience can also be viewed not only as a shattering of one's assumptive world but also as a shattering of one's manner of dwelling resonantly in the world. Positively, the anarchic care of parents creates a space for children to actualize their capacities to organize resonant experiences of being-in-themselves and being-for-themselves-with-others, which accompany a sense of self-efficacy. By contrast, early relational deprivations and impingements by parents (or others) heighten distrust and anxiety and undermine children's semiotic organizations of embodied being-in-itself and being-for-itself in relation to the environment. Put another way, traumatized infants are not simply distrustful of the environment; they cannot rely on their sense of being-in-themselves because they are under threat. Instead of experiences of resonant dwelling together, there is dissonance and a loss of self-efficacy. What we typically diagnose in adulthood as a narcissistic personality is someone who, early in life, has been deprived of the affection and attention necessary to develop a secure sense of being-in-itself and being-for-itself, which is necessary for a secure sense of resonant dwelling in the world. The metaphor of "emptiness" lies at the base of narcissistic wounds and is better understood as a profound sense of global embodied insecurity about being-in-itself and being-for-itself (later, insecurity in relation to senses of self-esteem, self-confidence, and self-respect). Indeed, the later development of being-for-and-with-others is disrupted because the narcissistic person seeks to use others instrumentally to secure their sense of being-in-itself, which only serves to screen fear or terror of vulnerability and a profound, pervasive insecurity. An adult who has been narcissistically wounded lives a dissonant life, frantically and unsuccessfully seeking experiences of resonant dwelling with others, though by turning "the world into a point of aggression … [the world] appears to us as something to be known, exploited, attained, appropriated, mastered, and controlled" (Rosa, 2020, p. 14). The metaphor of "emptiness" can be further understood as living in a world that lacks resonance. By contrast, one's good-enough assumptive world is a world of resonant dwelling and belonging, which, in early life, is understood as senses of being-in-itself and being-for-itself in relation to a trusted, caring parent who recognizes the alterity and affirms the singularity of the child. Deprivation or impingement, in brief, undermines infants' semiotic organizations of resonant dwelling and belonging in the world.

It is important to make a distinction between trauma and dissonance regarding one's assumptive world. For all kinds of reasons, human beings experience dissonance internally and in relation to others. Dissonance is not in and of itself connected to suffering or trauma. The dissonance an infant experiences is in relation to a parent who, for the moment, is preoccupied and fails to attune to the infant's assertions. This is disruptive but only moves in the direction of suffering and trauma if the relation is not repaired. A person who has an illness experiences dissonance that falls under the heading of suffering. Trauma, by contrast, is an overwhelming dissonance in one's assumptive world of resonant, embodied, relational experiences of being-for-and-with-others.

Other features implicit in one's semiotic assumptive world are agency and intentionality. Trauma that shatters one's assumptive world is accompanied by an undermining of agency and self-efficacy and an attendant sense of relative freedom. Trauma, in other words, represents the foreclosure of relational care and the corresponding collapse of the space of appearances—a space of anarchic care necessary for one to actualize one's potentialities, which include agency and freedom. In trauma, one has only one option for making use of one's agency, and that is sheer survival—bare life. The aftermath of traumatic experiences accompanies distortions in agency and intentionality because survival becomes the dominant aim of one's assumptive world, foreclosing the actualization of potentialities aimed at flourishing. Survival is, of course, important, but resonant dwelling with others and relative freedom entail much more than bare life.

Agency in the face of trauma can be further depicted in terms of Hartmut Rosa's (2020) interrelated notions of being affected and self-efficacy with regard to experiences of resonance. Our experience of resonance with others depends on being moved by the other. In being moved, affectively and physically, one experiences a sense of being *addressed* as a singular being or person (p. 32). This also accompanies self-efficacy, which is not only the sense of being connected or belonging but also the experience of having an effect on the other and the world. In trauma there is no sense of being addressed as a person—as a singular being. Rather, one is a mere mute object. A mute object, lacking singularity, has no self-efficacy because, despite one's cries, the victimizer does not respond, except perhaps to stifle the cries. Self-efficacy implies some measure of control and being addressed, but in trauma there is only the horror of absolute uncontrollability—the eclipse of singularity and self-efficacy.[6] This experience of uncontrollability becomes equated with vulnerability, which explains, in part, a person's resistance to the anarchic care of the therapist.

In Chapter 2, I mentioned that one's assumptive world of dwelling includes agency's capacity for impotentiality, which I include under the heading of self-efficacy. Recall Agamben's (1999) view that impotentiality is the capacity to actualize one's agency to prefer not to actualize one's potential to do something. As Agamben argues, "To be free is … *to be capable of one's own impotentiality*" (p. 183, emphasis in original). Put differently, an aspect of human dwelling in the world is to choose or decide not to actualize or live out some feature of one's assumptive world. In trauma, one is helpless to actualize one's capacity for impotentiality. One cannot act to render inoperative the violence and deprivation of *being* traumatized. Put differently, the very foundation of one's dwelling in the world—the core of one's assumptive world—is the belief and confidence in one's agency, which includes the *freedom to prefer not*. The shattering of one's assumptive world is the shattering of the belief in one's ability to exercise impotentiality.

Before continuing, an illustration is helpful. Jean Amery, a French journalist and resistance fighter, was brutally bound and beaten by the Gestapo during World War II. Some years later, he wrote about his experience of torture:

> The first blow brings home to the prisoner that he is helpless. … Yet I am certain that with the very first blow that descends on him he loses something we will perhaps call "trust in the world." Trust in the world includes all sorts of things … the certainty that by reason of written or unwritten social contracts the other person will respect my physical, and with it also my metaphysical, being. The boundaries of my body are also the boundaries of myself. My skin surface shields me against the external world. If I am to have trust, I must feel on it only what I *want* to feel. … The expectation of help, the certainty of help, is indeed one of the fundamental experiences of human beings. … The expectation of help is as much a constitutional psychic element as is the struggle for existence. … But with the first blow of the policeman's fist, against which there can be no defense and which no helping hand will ward off, a part of our life ends and it can never again be revived. … Whoever has succumbed to torture can no longer feel at home in the world.
>
> (Amery, 1995, pp. 126, 127, 136)

The first blow from the policeman represents a moment of profound depersonalization, wherein the singularity of Amery is absolutely denied. He is simply a mere object, an it, used to gain information. The policeman's fist also represents the point where Jean Amery was at his most vulnerable and the concomitant shattering of his assumptive world—a world of self–body integrity, of being-in-himself and being-for-himself, of trust in terms of the expectation of help and care, of ownership of his body, of self-efficacy, and of caring repair. It is likewise a moment of catastrophic loneliness and dissonance for no one responded to his cries for help, which entailed the complete absence of care and, of course, trust and fidelity. At the same time, Amery, in being helpless, was absolutely unable to render inoperative the grammar of torture—an absence of his agency and freedom to exercise impotentiality—in a word, uncontrollability. In his being tortured, his assumptive world no longer made sense, leaving these unbearable experiences unassimilable, unclaimed. The world after torture is a world in which he is not addressed and does not belong. It is a world where his agency is captive to mere survival—bare life. It is a world where dwelling is radically dissonant and *unheimlich*—home that is not home. It is a world devoid of resonance between human beings.

One may immediately question this because Amery later wrote powerfully about his experience, suggesting that he was able to claim and assimilate the experience into symbols and autobiographical memory. This is true, but only to an extent. Many people who have been traumatized can, in time, talk about their experiences (see Blanchot, 1995), but this does not mean that these unbearable, dissonant experiences are entirely claimed or worked through. In one sense, they cannot be claimed because the trauma, especially trauma at the hands of other human beings, is completely at odds with the assumptive, embodied, pre-representational world of trust and vulnerability—resonance. Put another way, experiences associated with severe trauma essentially lie outside human capacity for symbolization *because the very capacity for symbolization is existentially dependent on care, trust, and vulnerability*—on the openness to receive care (LaMothe, 1999). In brief, trauma is antithetical to personalizing care and trust, which are essential for the agency (and impotentiality), embodied senses of self-esteem, confidence, and respect that undergird our assumptive worlds.

Amery never fully recovered from his experiences of being tortured. He survived, for a time, and narrated his experience, but he lived in a world where he was alienated and isolated from routine meanings, purposes, and intimacies. His agency, in other words, was haunted by the specter of the trauma of having been tortured—a complete annihilation of resonant belonging and self-efficacy. As Fassin and Rechtman (2009) noted, "Trauma … is not simply the consequence of unbearable experiences, but [is] also in itself a testimony" (p. 20). And the testimony of Amery "is both the product of an experience of inhumanity and the proof of the humanity of those who have endured it [alone]" (p. 20). Victims of trauma carry a sense that this is a world where they no longer feel at home, which accompanies a profound sense of loneliness and dissonance.

I add here that Amery's trauma occurred when he was an adult, though I am suggesting that earlier foundations of his assumptive world were shattered. It is not simply that his early senses of being-in-himself and being-for-himself were undermined in the moment of trauma; it was also his resonant sense of being-for-and-with-others. As noted in the previous chapter, experiences of dwelling, developmentally, eventually include children's experiences of being-for-and-with-others qua persons. For Amery, torture and the isolation of his cell were experiences of catastrophic loneliness, which, in more abstract terms, is the radical absence of resonant experiences of being-for-and-with-others. Catastrophic loneliness is a manner of dwelling in a world where one does not feel existentially secure that one belongs with others in the world. What remains is an existential sense of dissonance and, correspondingly, the lack of hope of obtaining or recovering resonance.

Over the years, I have worked with courageous people who have been traumatized as children or adults. As a result of trauma, they no longer feel at home in their bodies, with others, or in the world. There is no resonance in dwelling with others, only the dissonance of being unhoused. For victims to face and delve into unclaimed experiences and a sense of catastrophic loneliness and dissonance with another human being, who, in part, serves as a witness to the trauma, takes an incredible amount of courage. This is the courage not only to remember the horrors but to risk vulnerability and uncontrollability with another human being who must, time and time again, earn their trust by consistently recognizing and caring for their singularity. It also takes courage to hope that they can resonantly dwell and belong in this world, if not as previously with an unquestioned secure sense of dwelling they experienced prior to the shattering of their assumptive world, then with a good-enough sense of being-for-and-with-others. It will be a world where they have a greater sense of their own agency and efficacy (and attendant senses of self-esteem, self-confidence, and self-respect) to actualize their potentialities (and impotentiality) for flourishing, even as the stain of trauma lingers in their psyches. The stain of trauma is the conscious or unconscious memory of the total collapse of the space wherein potentialities can be actualized. In brief, catastrophic loneliness is a memory of being unhoused from resonant experiences of dwelling and belonging in the world.

So far, I have been discussing trauma as it relates to individuals. There is also trauma experienced by groups or communities. Primo Levi, a Holocaust survivor, poignantly narrated his traumatic experiences and their close connection between personal and cultural objects and a sense of self. He wrote,

> But consider what value, what meaning is enclosed even in the smallest of our daily habits, in the hundred possessions which even the poorest

> beggar owns: a handkerchief, an old letter, the photo of a cherished person. These things are part of us, almost like *limbs of our body* [emphasis mine]. … Imagine now a man who is deprived of everyone he loves, and at the same time of his house, his habits, his clothes, in short everything he possesses: he will be a *hollow* [emphasis in original] man … for he who loses all often easily loses himself.
>
> (Levi, 1960, p. 27)

Levi is obviously talking about himself, but the backdrop of this is the brutal deprivation of Jewish cultural and religious symbols and practices that provided shared meanings and purposes necessary for mutual intimacies—for mutual experiences of resonant dwelling. As Terrence Des Pres (1976) writes, "Gone were the myths and institutions, the symbols and technologies which in normal times allow the self to transcend and lose sight of its actual situation" (p. 188). Stripping these away deprives individuals and a people of their capacity to make sense of their individual and collective singularity (shared senses of esteem, confidence, and respect) and, correspondingly, their making sense of the world. It is a world where previous symbols and rituals not only no longer make sense; they also fail to provide a traumatized people with a shared sense of belonging in the world. It is a collective shattering of assumptive worlds and unbearable experiences. It is a world of bare life.

The notion of catastrophic loneliness is applicable in collective trauma for two reasons. First, Levi's trauma and experience as a hollow man are personal. They are his experience of catastrophic loneliness. Even in the midst of other Jews who were traumatized, Levi felt alone and lonely. Recall that a central feature of our assumptive semiotic world in early life is that its very foundation is dependent on good-enough personalizing care. This becomes the unquestioned background of later experiences of dwelling. Trauma shatters the belief that the world is a place where one can dwell. In my view, when Levi was with others who were traumatized, his sense of catastrophic loneliness remained. Trauma isolates and alienates, even in the midst of shared experiences of victimization. Second, a group of people who have been traumatized can experience a collective sense of loneliness in relation to the larger world. Levi and numerous other European Jews felt abandoned by the world. Their cries and needs were overlooked, denied, or simply dismissed by other peoples, much like the Palestinian people today.

Another illustration of this is the treatment of native peoples by white European and U.S. colonizers. Jonathan Lear (2006) described how Plenty Coups, the Crow chief, led his people through a period of cultural devastation.[7] For Lear, cultural devastation involves the traumatic loss of narratives and rituals that a group of people uses to interpret current and past events, thus providing intersubjective meaning and purpose as well as a collective sense of singularity. These narratives and rituals make it possible to assimilate experiences, and, when these narratives and rituals are stripped away, making sense of the world and experience is undermined. For the Crow community, cultural devastation involved the violent and traumatic incursion of white Europeans into their lands, the dominance of white European narratives, and the corresponding loss of Crow cultural rituals and narratives that had provided meaning, a sense of hope, purpose, and direction for a Crow future. Myths, shared stories, and rituals, in other words, provide the ground for a shared and relatively secure and resonant sense of dwelling in the world. The stripping away of these, as experienced by the Crow people as well as by millions of Africans who were violently enslaved by European colonizers, undermines the assumptive worlds of shared belonging and dwelling. It may not be catastrophic loneliness for the Crow people, though I think it was in the sense of other peoples (e.g., settlers) contributing to and not recognizing the sufferings of the Crow people. In brief, the deprivations were a catastrophic, existential semiotic crisis wherein their assumptive world of dwelling was radically undermined. Individual and collective trauma, in short, can be understood as an overwhelming disruption of the semiotics of resonant dwelling—an assumptive world of being-in-itself, being-for-itself, and being-for-and-with singular others.

For the sake of further clarification, I want to shift the discussion about trauma to the reality of existential insignificance. In Chapter 2, I indicated that all semiosis exists in relation to existential insignificance and the impermanence of our significations. Death is the cessation of life and semiosis. In good-enough parent–child relations, the anarchic care of parents provides spaces for children to gain a sense of their own agency in constructing experience, as well as confidence and security in their sense of going on being (being-in-itself, being-for-itself) against the background of existential insignificance and impermanence. If children are forced, violently or otherwise, to encounter this reality, I would argue it is traumatic. First of all, they are not *choosing* to face or accept existential insignificance or the impermanence of their own assumptive world. Second, in the moment of trauma, they are alone in facing this abyss. Children can face their emerging awareness of death and existential insignificance if accompanied by loving people who provide significance in the face of existential insignificance. A child's mother dies, and the child is consoled by the father and others, reassuring the child in different ways that significance (the child's own and the mother's) will continue in the face of this death. We also do this as adults when facing the death of loved ones. We find all kinds of ways to ensure the person's significance (and our own) in the face of death (rituals, grave markers, memorials, etc.). In trauma, one is forcibly faced not simply with the absence of one's own significance (trauma as depersonalization or denial of singularity) but also with the stark reality of existential insignificance without the reassurance from others of one's own significance. In other words, an individual or group, in the midst of trauma, is forcibly stripped of transitional or other objects that provide assurance of their significance in the face of existential insignificance. This can also be understood as catastrophic loneliness and dissonant dwelling.

This may raise a question regarding whether facing existential insignificance is not in itself traumatic. Using Donald Winnicott as an illustration, I do not think it is, except in situations of being forced to face the abyss alone. In 1954, Winnicott suffered his first heart attack in his middle age. Fourteen years later, he had a severe cardiopulmonary crisis, three years before his death (Rodman, 2003). To understand Winnicott's experience of and response to the reality of his impending death, I must first turn to his view of early infancy. During infancy's period of relative dependence, according to Winnicott (1965),

> the ego changes over from an unintegrated state to a structured integration, and so the infant becomes able to experience anxiety associated with disintegration. The word disintegration begins to have a meaning which it did not possess before ego integration became a fact. In healthy development at this stage the infant retains the capacity for re-experiencing unintegrated states, but this depends on the continuation of reliable maternal care or on the build-up in the infant of memories of maternal care beginning gradually to be perceived as such. The result of healthy progress in the infant's development during this stage is that he attains to what might be called "unit status". The infant becomes a person, an individual in his own right.
>
> (p. 44)

Here we note that, in a good-enough environment, the child's emerging agentic capacity to organize experience takes place in relation to unintegration.[8] In my terms, the parent's anarchic care enables the infant to experience and tolerate, manage and make creative use of unintegrated states. Indeed, for Winnicott, unintegration is a source of creativity and health.

The term "unintegrated states" is worth saying more about before we proceed. From my perspective, unintegrated states imply experiences that are not integrated by the ego.[9] Winnicott associates these with the unconscious. There is some truth to this, but the notion of the unconscious already indicates the capacity for semiosis. To integrate what is unintegrated already implies some unconscious, semiotic organizations of experience. Yet, the emergence of a nascent ego and semiotic organizations of experience takes place against the background of existential insignificance, which is the ultimate "unintegration." There are two senses of unintegration, then. The first is unintegration in terms of cognitions and affects that are not yet integrated as conscious experiences, and these, often presymbolic, can evoke anxiety. In other words, it is certainly feasible that a child has some anxiety when facing unintegrated states, in part because they challenge the integrated states of the ego. However, there is a more profound existential anxiety, and that is the reality of existential insignificance. Existential insignificance is unintegration in the absolute sense. It is the absence of semiosis, whether we are speaking about consciousness or the unconscious. Existential insignificance cannot be captured by semiotics. Semiotics can, like Dickens's (2013) Ghost of Christmas Future, merely point to it, but it cannot be experienced, which is why Freud said we cannot experience our own death. This said, parents' anarchic care, which facilitates the actualization of children's egos and conscious and unconscious (unintegrated) semiotic organizations, provides solace, significance, and resonant belongingness in the face of existential insignificance. Anarchic care, then, is a semiotic relational activity that makes it possible for persons to feel alive and real in the face of anxieties related to unintegrated experiences as well as existential insignificance or ultimate unintegration.

In facing death, Winnicott turned to poetry—an expression of agency and creativity in the face of the ultimate loss of agency, separation, and, for him, unintegration. Winnicott knew he was dying and again he made use of the transitional phenomena of poetry:

> Let down your tap root
> to the centre of your soul
> Suck up the sap
> from the infinite source
> of your unconscious
> And
> Be evergreen.
> (as cited in Kahr, 1996, p. 123)

In this poem, we note his embrace of unintegration—for him, the infinite source of the unconscious semiotic organizations of experience. The "soul" as well represents something that is not able to be fully grasped yet is known. At the same time, it is important to recall the context of the poem—Winnicott's age and his severe cardiac crisis. Stated in a nonpoetic way, Winnicott was using his agency to integrate what cannot be integrated—not knowing and death. If we allow for the context, it becomes clear that the poem signifies Winnicott's experience of going on being, and that this foundational experience of going on being was enlivening and creative (evergreen), even in the face of his impending death. This poem represents the agentic play of creative illusions (let down your taproot—your agency) amid real experiences of being alive in the face of terminal, absolute unintegration. Stated differently, the poem signifies the aliveness of a "true" self in facing what he deemed to be unintegration but what I am calling the reality of existential insignificance.

He made a similar and very telling comment in a lecture during the last year of his life. Winnicott told the audience, "A great deal of growing is growing downwards. If I live long enough I hope I may dwindle and become small enough to get through the little hole called dying" (as quoted in Kahr, 1996, p. 125). Is this not a reversal of his comment about birth? A nascent ego or agency is involved in the baby's struggles to be birthed into a new reality—one that is initially formless or in a state of relative unintegration. The baby, while helpless, to a small degree participates in their birth—a birth experience that they are unable to integrate. It is this small ego that cooperates in the movement to life outside the womb. Winnicott believed that, at the end of life, the ego, which has grown considerably, ideally needs to become smaller so it can go through the little aperture that leads to formlessness, to the unknown, to unintegration. Winnicott was, I believe, saying that to accept the infinite—that which cannot be integrated—one must choose or act to render the ego smaller. One notes as well that this comment and the poem above are free of anger, resentment, resignation, or heroic fighting in the face of dying and death. The absence of fighting or strident agency does not connote passivity. Rather, Winnicott manifested an active, agentic, self-efficacious, creative participation in the face of the ultimate formlessness—the reality of death. In other words, his participation is the paradoxical act toward making the ego smaller—through agency—in the face of the helplessness and absolute dependency experienced in dying. In terms of helplessness or uncontrollability, Winnicott could do nothing about the reality of his dying and his eventual death. He could not exercise, by way of agency and efficacy, his impotentiality and render death inoperative. Death is uncontrollable, but we have, as Winnicott's poem exemplifies, some measure of control in how we face the uncontrollable (Rosa, 2020). While he could not render the reality of death inoperative, he used his agency to participate in the dwindling of his ego. Perhaps the illusion is that one indeed makes an ego or agency small enough to fit through the tiny hole of death—a hole that cannot be rendered inoperative.

Another important example of Winnicott's attitude and experience of helplessness is seen in a religious comment he made toward the end of his life. In his unfinished autobiography, he wrote a prayer: "Oh God. May I be alive when I die" (as quoted in Kahr, 1996, p. 125). First of all, note that Winnicott's prayer is not for eternal life. There is no illusory belief about eternal life when facing the reality of death, yet there is the illusory belief in the object of his prayer—a transitional object. Further, the petition reflects Winnicott's helplessness—something he had no absolute control over—and absolute dependency in the face of death. And yet, the prayer represents an ego-integrating experience in the face not of primary unintegration (a semiotic experience not yet integrated into conscious formulations) but of terminal unintegration. To be and feel alive and real (an experience of singularity) at the time of one's death represents semiotic experiences in the face of existential insignificance. This prayer must also be seen in light of the last year or so of Winnicott's life. During the final 18 months of his life, Winnicott made a point of saying good-bye to friends and colleagues, even as he continued to work. All of this points to a man who embraced the great unintegrating realities of human life—dying and death—while still being creative and alive with other people.

Winnicott's prayer indicates a courageous willingness to accept (agency) unintegration, the unknown, the infinite, or, in my view, existential insignificance.

All of this said, there is some ambiguity in Winnicott's poems about death in light of his terms of integration and unintegration. As noted above, if unintegration refers to states outside of the ego's organizations, then semiosis continues to be applicable. The unconscious is by definition part of semiotic activity. Do death and the infinite simply represent absolute unintegration? Is his use of the terms "soul" and "evergreen" meant to convey something that transcends the ego and death? To my mind, it is better to leave the terms "integration" and "unintegration" to conscious and unconscious semiosis. It is clearer to me to replace unintegration with existential insignificance, which represents the absence of both life and semiosis.

Now let me explain why Winnicott's facing existential insignificance was not traumatic. In facing his dying and eventual death, Winnicott wrote and gave lectures, he penned poems, and he engaged in caring and loving relationships with friends and colleagues. In the face of anxieties and uncertainties regarding his impending death, Winnicott was able, in numerous ways, to exercise his signifying capacities and agency, providing experiences of efficacy, of resonant belonging with others, of being alive and real even as he faced the reality of existential insignificance and impermanence. Perhaps finishing a collection of his works was his attempt to establish a legacy that would continue after his death, but this semiotic activity (establishing a legacy) is illusory in the face of existential insignificance. Regardless, Winnicott's facing existential insignificance was not a shattering of his assumptive world or an experience of catastrophic loneliness and dissonance, though I imagine he did experience a sense of loneliness in facing death. Nevertheless, he retained his assumptive world while dying and facing imminent death.

There is, however, trauma associated with existential insignificance. This occurs when an individual's assumptive world is forcibly shattered and the individual is driven to face the reality of existential insignificance alone—catastrophic loneliness. To explicate further, ideally speaking, a child's assumptive world is thoroughly agential, semiotic, relational, and founded on anarchic care over and against the backdrop of existential insignificance. A child may face the death of a parent, which is painful, distressing, and disturbing. It is not traumatic, as noted above, if the child is comforted by other loving adults who, through their love, assure the child of their ongoing significance and attendant assumptive world in the midst of the precarity of existential insignificance. Trauma, however, entails depersonalization (denial of singularity, of being-in-itself and being-for-itself) and the absence of care. Both depersonalization and the absence of care represent the undermining of the very foundational features of one's assumptive world. The shattering of one's assumptive world (the violation of resonant being-in-itself, being-for-itself, being-for-and-with-others) means being forcibly exposed—without recourse to the comfort provided by one's assumptive world—to existential insignificance alone. Remember, one's assumptive world is a semiotic construction dependent on anarchic, personalizing care in relation to the reality of existential insignificance. This assumptive world provides individuals with a sense of agency, self-esteem, self-confidence, and self-respect, all of which provide solace and significance in relation to the background reality of existential insignificance. To recall Jean Amery and the policeman's fist that shattered Amery's assumptive world, it was not just Amery's body that was violated; his assumptive world was also violated and could no longer provide comfort or meaning in the face of depersonalization and the forcible facing of the abyss of existential insignificance—an abyss he faced alone and without the solace of his assumptive world.

Naturally, not all trauma entails a singular event or a dramatic event. Decades ago, Masud Khan (1963) used the term "cumulative trauma" to refer to ongoing deprivation. Many therapists have worked with patients who grew up being deprived of affection, attention, and anarchic care. To be sure, they received enough care to survive, but their assumptive world is insecure and rests on shame—consistently inadequate parental care leads to a profound sense of inadequacy connected to a diminished sense of self-esteem, self-confidence, and self-respect, all of which impact agency. They cannot point to a single event that haunts them or even to a series of events—just a background of carelessness, leaving them feeling dread, a sense of unreality, and deadness (Green, 1999). This is not so much a shattering of their assumptive worlds as an insufficiency of an assumptive world that supports and accompanies their sense of being alive in the face of existential insignificance. At their core is an insecurity, wherein resonant senses of being-in-itself, being-for-itself, and being-for-and-with-others are tenuous semiotic organizations. More colloquially, profound shame accompanies a pervasive sense of aloneness and loneliness, making it nearly impossible to face, let alone accept, their existential insignificance—mainly because they do not possess a reliable and secure sense of their own significance. Put differently, this loneliness is also understood as not feeling at home in the world with others.

An illustration can help anchor some of the main points above. Zoe[10] was 38 when we first met. She was referred by a friend who knew about her situation at home. Her husband of 10 years was an alcoholic and was emotionally abusive to her and their two children. Zoe was an exceptionally intelligent, successful professional who, for all practical purposes, was a single parent. What triggered her motivation to seek therapy was a medical scare (possible breast cancer) three months previously. Inexplicably to her, Zoe began having flashbacks, feelings, and thoughts that were overwhelming. At one point, Zoe said that she had kept those experiences in a box but, after the surgery, she no longer could keep them there. Over a period of months, Zoe slowly, in fits and starts, began to tell me about those experiences. Around the age of 7, her father began to sexually abuse her—abuse that escalated into violence. Her mother was an alcoholic, and Zoe said she was mostly mentally or emotionally absent, which meant Zoe lost any hope of having her mother step in to protect her. Nevertheless, Zoe was very protective of her mother (mother as the sole good object), to the degree that she could be. As she entered puberty, Zoe sought shelter by staying at friends' houses, though on a couple of occasions she was sexually abused at those homes. The abuse ended when Zoe left home after high school, with stints on the street. Amazingly, she was able, through grit, hard work, and creative determination, to obtain a college degree. Nevertheless, she was haunted by these memories, never telling anyone because she felt so much shame and helplessness. She also feared that these memories might escape the psychic box she had created to survive.

Zoe's assumptive world was shattered when her father abused her (and her mother failed to protect her). Zoe's father's abuse represents actions devoid of personal recognition and care, undermining her sense of being-in-herself, being-for-herself, and being-for-and-with-others. Put another way, Zoe, in those moments, was a mere object to be used—an object of no significance, no singularity. As she was a mere object, this meant that her previous senses of being-in-itself, being-for-itself, and being-for-and-with-others were violated. That is, the abuse, which was grounded in the father's depersonalization of his daughter, represented the collapse of a space wherein she could actualize senses of self-esteem, self-confidence, and self-respect—senses that undergird agency, self-efficacy, and embodied, relational resonance. In terms of agency, Zoe's experience of helplessness meant she was unable to actualize her capacity for preferring not or rendering the abuse inoperative. Indeed, her subsequent defenses (putting the memories in a box) can be understood as her being unable to render the memories inoperative, which only exacerbated her feeling of helplessness. Instead of an agency connected with a sense of resonant dwelling and belonging with others and a capacity for inoperativity, Zoe's agency was focused on sheer survival in a world (and home) where she was catastrophically alone and lonely. Her courageous work in therapy (and elsewhere) over several years eventually made it possible for her to render these memories inoperative—meaning she was able to put the past in the past. In so doing, she experienced feeling less alone and lonely.

It was not simply that Zoe, after the first experience of abuse and the shattering of her assumptive world, no longer felt at home in her family's house. As in Amery's case, no one responded to her pleas for help, and no one recognized her suffering after the abuse. The absence of help in the context of depersonalization became linked to a belief in and experience of worthlessness—a self without singularity, without existential value. Overwhelming shame is an experience dominated by the belief that one is worthless—wholly insignificant—which represents profound dissonance. This shame, which she tried to keep hidden from herself and others, was part of the reason for her refusal to trust, to be vulnerable in relation to others. She was desperately afraid that people would see her and inevitably reject her, which exacerbated her loneliness. Put another way, shame represents not only internal dissonance but also a relational dissonance in dwelling with others. All of this was joined to the illusion of invulnerability, which was represented by the "box" that, after surgery, was no longer able to contain her memories of trauma and shame. This illusory invulnerability kept her in the cage of feeling lonely in her marriage, with her friends, and in relation to her children. When she married, she hoped (or, better, wished) that she would not be lonely and would feel at home with her husband, but holding on to the illusion of invulnerability ensured the continuance of loneliness. When she had children, Zoe again felt hope in the possibility of dwelling, of being-at-home with herself and her children. While she was fiercely protective of her children, her invulnerability kept alive her profound senses of loneliness and dissonance (embodied and relational).

From a different angle, part of Zoe's initial assumptive world was the belief in her significance (singularity) that screened and unconsciously provided comfort in the face of the reality of existential insignificance. The abuse stripped her of her assumptive world of significance or singularity, forcing her to face existential insignificance alone. Her helplessness meant Zoe could not choose to make her ego smaller or to act creatively in relation to unintegration because unintegration was linked to unclaimed traumatic experiences. Worse, unlike other people who face existential insignificance with others who provide solace, care, and support, Zoe was absolutely alone, relying on psychological defenses to survive—a bare life. Beginning therapy was her first step in a long, arduous trek with someone who could witness her suffering, recognize and care for her singularity, and accompany her in facing the existential reality of the impermanence of significance.

I have a final point to make about trauma. So far, I have been discussing trauma as an event(s) that shatters persons' assumptive worlds. Can a future event, which is not experienced, be traumatic in the present? Can a future event shatter one's assumptive world? These questions emerge against the background of the climate polycrisis and the growing awareness of the numerous future disasters human beings (and other species) will experience (e.g., massive floods and fires; loss of large tracts of arable land [desertification]; hugely destructive storms; the collapse of insect, bird, and fish populations [and the starvation of human beings]; increasing global conflicts; failed states; and mass movements of peoples). While I do not think a future possible event(s) can lead to the shattering of one's assumptive world in the present, I do believe eco-emotions such as anxiety, fear, despair, remorse, and sadness (Cunsolo, Borish, et al., 2020; Cunsolo & Landman, 2017; Engstrom, 2019; Pihkala, 2018b, 2020a, 2022a, 2022b; Wu et al., 2020) anticipate traumas related to dwelling in the world. Like the canary in the coal mine that alerts miners to imminent disaster not yet experienced, eco-emotions alert us to the traumas that are on the horizon, our own and those of other species. Will the alert be enough to motivate us to take action to avert these traumas?

In summary, trauma, which occurs at an extreme point of vulnerability, is a massive disruption of a person's semiotic (symbolic) capacities associated with their assumptive world. These unclaimed experiences, which the individual is unable to render inoperative, accompany catastrophic loneliness and, correspondingly, an enormous disturbance in their sense of dwelling in the world with others. In other words, since experiences of dwelling depend on caring interpersonal relations, human trauma represents the eclipse of personal recognition and care, resulting in (1) a profound sense of being unhoused, (2) an undermining of one's resonant senses of being-in-itself, being-for-oneself, and being-for-and-with-others (dwelling together), and (3) an eclipse of one's associated capacities for agency, self-efficacy, and freedom. Finally, catastrophic loneliness in the shattering of one's assumptive world of resonant dwelling can also be understood as a consequence of one's being forced to face the abyss of existential insignificance without the solace of one's assumptive world resonantly shared with others.

Trauma and the Silence of Other Species

To focus first on human beings is not to privilege them, at least not from my perspective. Rather, it reflects an understandable moral commitment or imperative to recognize, understand, and respond to the suffering of singular human beings who dwell in the world with us. And yet, our dwelling, as I have labored to point out, also includes other species that share the same world. Since all life is semiotic, all living beings dwell in the world (being-in-itself and being-for-itself in the world). To dwell, then, necessarily implies singularity, which raises questions regarding the suffering and well-being of a singular being (Birch, 2024; Nussbaum, 2022; Singer, 1975). Put differently, acknowledgment of singularity necessarily moves one into the deliberative realms of care and justice. Yet, by all accounts, the ontological rift and its attendant normative, historical, and political unconscious lead to our overlooking not only the sufferings of other species but also our reluctance to consider whether these sufferings fall under the heading of trauma. I suspect that many human beings will accept that sentient animals, such as our cats and dogs, can experience a shattering of their assumptive worlds, yet, at the same time, most of us overlook the traumas of the sentient animals that we consume, such as cattle, pigs, and sheep. What about species that are invertebrates? Are they sentient? Do they suffer? Can they be traumatized? Do insects suffer? Can they be traumatized?

These are complex questions that cannot be addressed here, though I believe, if we are to bridge the rift, we must be willing to engage in philosophical-ethical discourses involving the sufferings and traumas of other species. My premise is that all living beings possess various types of semiosis and sentience, varying in complexity. Put another way, all semiosis entails an assumptive world necessary for survival and flourishing. This assumptive semiotic world can be crushed while the singular being continues to live. I add here that philosophers[11] and others have long recognized that sentient creatures suffer, which requires our finding ways to alleviate or stop unnecessary suffering. Yet, suffering, as noted above, is not identical to trauma. Below I use the imaginative story by Jack London regarding the life of Buck to illustrate his trauma at the hands of men and his healing that was the result of the reliable care of a man who rescued him. My premise is that any living creature, since it possesses the capacity for semiosis necessary to dwell in the world, can have its assumptive world shattered. The challenge lies in our willingness and ability to do the work to understand not only a precocial or altricial creature's assumptive world but also the creature's sufferings and traumas.

Before continuing, I wish to state again that the issue of trauma of other species is an analytic concern because psychoanalysis is concerned about suffering and trauma, as well as about raising to consciousness what we have repressed or dissociated—individually or collectively—as a result of the ontological rift. Culture, society, politics, and economics can all be put on the couch with the aim of facing what we have ignored out of fear or anxiety. Jacqueline Rose writes,

> It is a central tenet of psychoanalysis that if we can tolerate what is most disorienting—disillusioning—about our own unconscious, we are less likely to act on it, less inclined to strike out in a desperate attempt to assign the horrors of the world to someone, or somewhere else. It is not, therefore, the impulse that is dangerous, but the ruthlessness of our attempts to be rid of it.
>
> (as quoted in Morton, 2024, p. 53)

Once we are mindful of what was previously unconscious, we are faced with a decision regarding what to do with it.[12] We are responsible, though we can ignore our responsibility or rationalize and justify our actions, in this case related to the unnecessary suffering of other species. For instance, we can rationalize or justify our purchases of meats associated with industrial slaughter factories. Rationalization and denial do not abrogate our responsibility in participating in the apparatuses that cause untold sufferings and traumas.

There is another analytic reason for understanding the traumas of other species. As stated in the introduction, psychoanalysis is concerned with psychosocial development, which includes human flourishing. The climate polycrisis is leading to a degraded Earth and the loss of biodiversity, which is undermining and will continue to undermine human well-being and dwelling, as well as lead to widespread traumas—human and other-than-human. We can ignore discourses concerning ethics or justice regarding other-than-human species and simply be instrumentally and narcissistically interested in maintaining biodiversity by caring for other species. In any case, psychoanalysis can play a role in heightening awareness and self-reflection regarding what the ontological rift has repressed or dissociated. Of course, my main interest is in placing instrumental relations in the background while foregrounding our ability to understand and ethically respond to the sufferings of human and other-than-human beings.

Now let me turn to an imaginative tale about trauma. Jack London's (1903/1981) *Call of the Wild* begins in Santa Clara, California. Buck lived with Judge Miller and, as London narrates, "Buck was neither house-dog nor kennel-dog. The whole realm was his" (p. 762). We can imagine that Buck's initial assumptive world entailed mutual care and shared resonant dwelling with the judge and his family. In this bucolic setting, there is an air of foreboding. Unbeknownst to Buck, "Men, groping in the Arctic, had found yellow metal [and] these men wanted dogs" (p. 761). One of the judge's laborers, Manuel, while the family was away from the house, sold Buck to another man who transported dogs to Alaska Territory. Buck trusted Manuel, accepting "the rope in quiet dignity," unaware of Manuel's treachery (p. 763). When Buck growled after being handed over, the man tightened the rope, leaving Buck unconscious. Buck was then thrown into the back of a carriage. In captivity and on the long trek to Alaska, Buck was brutally beaten, painfully learning that the man with the club was the lawgiver (p. 768).[13] Buck's first day in Alaska

> was like a nightmare. Every hour was filled with shock and surprise. He had been suddenly jerked from the heart of civilization and flung into the heart of things primordial. … All was confusion and action, and every moment life and limb were in peril.
>
> (p. 770)

In the weeks and months ahead, Buck was used to pull a sled with other dogs, finding ways to survive brutal treatment, as well as competing with other dogs on the team for scarce resources. It was a bare existence or life.

The "law" of the club, which represents traumatic violence or a threat of violence toward Buck and the other dogs, signifies the dominion of the owners, the subjugation of Buck, and the instrumental relation that existed between them. After being sold to various men, Buck's last "owner," Hal, was a particularly brutal man. One time Buck, after an exhausting, cruel trip, was physically spent. Hal, the owner, was having none of it: "Get up there Buck! Hi! Get up there! Mush on!" (London, 1903/1981, p. 805). London wrote, "This was the first time Buck had failed" (p. 806). Hal then proceeded to beat Buck mercilessly and only stopped because a gold prospector named John Thornton, who had angrily observed this, intervened, knocking Hal aside. Thornton said, "If you strike that dog again, I will kill you" (p. 806). Hal whined that he owned the dog, yet he had "no fight left in him" (p. 807) and departed with his sister and a friend across the ice-covered river. John "knelt beside [Buck] and with rough, kindly hands searched for broken bones." His search "disclosed nothing more than many bruises and a state of terrible starvation" (p. 807).

As an aside, what happens next is, in my view, metaphorically prophetic regarding the trajectory of the ontological rift and the climate crisis. Hal, his friend (Charles), his sister (Mercedes), and the dogs made their way across the icy expanse.

> Suddenly, they saw the back end drop down, as into a rut, and the gee-pole, with Hal clinging to it, jerk in the air. Mercedes's scream came to their [Thornton's and Buck's] ears. They saw Charles turn and make one step to run back, and then the whole section of ice gave way and the dogs and human beings disappeared. A yawning hole was all that was to be seen.
>
> (London, 1903/1981, p. 808)

Hal's instrumental use and vicious treatment of Buck (and the other dogs) represent the presence and consequence of the ontological rift. In reading this part of the story in light of the climate emergency, it strikes me that we are not heading across a frozen expanse but rather are on a trajectory where millions of species and very likely humanity will drop into the yawning abyss of extinction—a consequence of many but not all human beings' callous instrumental treatment of other species and the Earth.

This detour aside, in terms of trauma, Buck's assumptive world was repeatedly shattered after Manuel sold him to the trader. The world of the judge and the homestead in California was, relatively speaking, one of resonant dwelling and belonging—freedom, care, trust, and acceptance. Using the terms above, one could say that Buck's assumptive world entailed a sense of being-in-himself, being-for-himself, and being-for-and-with-others (the judge's family), though I will qualify this below. Buck was at home with the judge and his family, which included workers such as Manuel. Buck's way of dwelling in the world was radically altered when he was betrayed, beaten, and later sold on the beach in Alaska. It was a world of brutal instrumental relations, wherein dogs were constructed as inferior (lacking singularity) to human beings and, therefore, available to be used and discarded when they no longer served a purpose. Put differently, Buck and the other dogs were instrumentally significant but, when they could no longer fulfill that function, they were deemed to be insignificant and thus easily abandoned. I add here that the brutality of the men who sold Buck and other dogs led to what we might call catastrophic loneliness. Even in the presence of other dogs, Buck was radically alone because he had to fight them to obtain the meager rations to survive. It was a kind of lonely, dissonant dwelling in a dangerous world.

It is important to situate Buck's traumas within a larger context because it helps us see that individual traumas emerge from, more often than not, macro systems and forces that function to legitimate violence while ignoring the singularities of those traumatized. In Alaska Territory (and in Canada), as

London pointed out, gold had been discovered, which set off a stampede of mostly men looking to get rich, whether through the discovery and mining of gold or by providing equipment and services to miners and others. Any gold rush at the end of the 19th century and the beginning of the 20th century could not and cannot be understood in the absence of capitalism and colonization. Indeed, during this period, capitalism's extractive, expropriative, and exploitive features were, in some ways, at their height because of a lack of regulations and protections. In terms of London's story, the extractive element of capitalism was evident in the mining of gold without concern for Alaska Native peoples or the environment in which they had lived for thousands of years. The expropriative feature was the taking possession of land from Indigenous peoples. The exploitive element was revealed in the earliest paragraphs of the story. Because of the market demand for large dogs in Alaska, men were stealing or purchasing dogs from outside the territory to maximize their profit. "Dogs" were mere cogs and, in that sense, all the same—existentially insignificant.[14] In terms of colonization, while there are Alaska Native peoples in London's story, they are peripheral actors. Toward the end of the story, the Yeehats (a local native community/nation) end up ambushing and killing John Thornton and his companions, Hans and Pete. This broke the last tie that held Buck to "civilization." The Yeehats, who are constructed as "uncivilized," function in the story to move the plot along. Whether intended or not, London did not highlight the plight and traumas of Indigenous peoples[15] as they sought to protect their land from the violent, traumatizing incursion of white European Americans and their attempts to expropriate Alaska Native peoples' lands.[16] Like in other tales of U.S. expansionism, the stories of Indigenous peoples are simply noted to advance the plot of U.S. colonization and imperialism.

Colonialism and capitalism are systems that arise from and reproduce the ontological rift, inflicting innumerable sufferings and traumas upon othered human beings and other species. To return to London's story, there is a human-created abyss between Buck and his "owners." Put differently, the assumptive world of the colonizer and the capitalist entails beliefs and perceptions that dogs like Buck are, at best, useful inferior objects that merit some degree of care as long as they continue to serve their masters well. The colonizer's assumptive world radically disconfirms Buck's assumptive world, and it does this through violent subjugation and depersonalization. These two worlds cannot abide together. Buck's initial assumptive world is one of being-in-himself, being-for-himself, and being-for-and-with-others. In this world, Buck possesses a sense of singularity in resonant relations with others. In the assumptive world of the colonizer and the capitalist, the object of sale possesses no singularity, though it has limited and impermanent use value. It is a world of horrific dissonance and trauma.

There are two interrelated points here. First, the assumptive worlds of the ontological rift are incompatible with assumptive worlds wherein the singularities of other species (and othered human beings) are recognized and affirmed. To genuinely recognize the singularity of Buck, for example, alters one's relationship to Buck. John Thornton, unlike Kantians, felt the categorical demand to intervene when he saw Buck's suffering. Thornton did not simply intervene to stop Hal's brutality; he also tended to Buck's wounds. There was no instrumental relationship in his anarchic care for Buck because he recognized Buck's singularity. Second, the assumptive worlds that produce the ontological rift may not be in themselves traumatizing, but they sow the seeds for exploitative, traumatizing relations. London was not aware of climate change and, likely, would not have identified colonial expansionism in terms of anything like the notion of the ontological rift. But this story of Buck is one that reveals the rift and its manifold traumas that result from human beings' construction of other species and othered human beings as lacking singularity, which, in turn, gives rise to forms of violent depersonalization.

Let me return to the story and the possibility of repairing the rift and the process of healing the trauma. Once Hal and company were out of the picture, John began to take care of Buck. John's care renders the apparatuses of capitalism and imperialism inoperative, creating a space that is devoid of instrumental relations (anarchic care). More positively, John's care opened up a space for Buck to experience his singularity. As Buck's wounds healed and he put on weight, he discovered, to his surprise, that the other dogs "manifested no jealousy toward him. They seemed to share the kindness and largeness of John Thornton" (London, 1903/1981, p. 808). Thornton genuinely cared for the animals in his company, and this care meant that Buck and his mates did not have to compete for sustenance to survive. The relations between the dogs and the men were characterized by cooperation and reciprocity. There was a resonance between them as they all dwelled together.

As a result of the consistent recognition and treatment of Buck's singularity, "Buck romped through his convalescence and into a new existence" (London, 1903/1981, p. 808). "Love," London continues,

> genuine passionate love, was his for the first time. This he had never experienced at Judge Miller's down in the sun-kissed Santa Clara Valley. With the Judge's sons, hunting and tramping, it had been a working partnership. … But love that was feverish and burning, that was adoration, that was madness, it had taken John Thornton to arouse.
>
> (p. 808)

Buck recognized that other men "saw to the welfare of their dogs from a sense of duty and business expediency; he [John] saw to the welfare of his as if they were his own children" (London, 1903/1981, p. 809). The mention of "children" suggests a kind of relation marked by the recognition of Buck's singularity, accompanied by a kind of anarchic care absent from instrumental epistemologies, beliefs in superiority and inferiority, and relations of dominion and subjugation evident in the ontological rift. More positively, John cared for Buck without any idea or aim that Buck would work for him—Buck was valued in himself. Buck, for instance, was free to remain in the camp or leave, which he eventually did. This unconditional care was accompanied by an unconditional trust, wherein John's trust in and loyalty to Buck was free of contractual expectations, use value, or splitting. There was, in this relationship, no normative or historical or political unconscious. Buck was part of John's story and vice versa. Buck's new assumptive world included resonant senses of being-in-himself, being-for-himself, and being-for-and-with-others. In brief, what is evident in this mutual affection is the absence of the ontological rift and the presence of the sense of resonant dwelling in the world with others.

Buck recognized that his original assumptive world with the judge, while obviously preferable to what happened at the hands of Hal and others, was radically different from the assumptive world he was learning to live in with John Thornton and the other dogs. This further clarifies my point above regarding the ontological rift and trauma. The judge and his family treated Buck well, yet it was nevertheless an instrumental relation, wherein Buck's significance and care were tied to his functioning as a family companion. There was no trauma, but the background of the ontological rift remained. Buck's epiphany regarding John's love, which entailed the recognition of Buck's singularity, was exactly the radical difference between the assumptive world of California and the assumptive world that was emerging with John.

London's story ends with John Thornton, his partner, and the other dogs brutally killed by Yeehat men. This ending can add to our understanding of trauma, not only of human beings but between human beings and other species. Buck, who had been spending more time in the "wild,"[17] happens upon the scene and becomes enraged. Again, his assumptive world collapsed, but this time he hunts down and kills the men responsible. With John dead, there is no hope that Buck will find love in the so-called civilized world. Instead, Buck joins other dog-wolves in the wild. There is a movement from civilization to nature in London's story, whereas, in another of his stories, *White Fang*, there is a movement from the wilds to civilization. As I noted in Chapter 1, the juxtaposition of civilization and nature is a product of the ontological rift.[18] While this does not mean that trauma will occur, I suggest that this relation represents the tragic repetitive compulsion of the rift. The rift is present in Buck's early years living on the judge's land. Buck is cared for, but not as a singular being. He was still being used, which only occurred to him after John's noninstrumental, anarchic care. Nevertheless, this early assumptive world was shattered shortly after Buck was sold. Despite the horrors and suffering in the shattering of his assumptive world, Buck found solace and some measure of healing in the anarchic care of John Thornton. The Yeehats, who almost certainly had had their world shattered by colonizers, took their revenge on John and his party. Death is the ultimate shattering of one's semiotic world. Buck experienced the rending of these deaths, but he did not mourn. Instead, the sufferings of past traumas and the present losses morphed into deadly rage as he hunted down the men responsible. Trauma often begets trauma, which is a vicious cycle born of the ontological rift. The ontological rift begets violence and trauma. John's anarchic, noninstrumental love for Buck points to the possibility of crossing the rift or rendering the rift inoperative, breaking this cycle of trauma, if only for a time.

Implications of London's Story for the Climate Polycrisis

In Chapter 1, I mentioned a conversation between two philosophers, one of whom said that her friends had given her a book that would convince her not to eat meat. The interviewer asked if she had read it, and she laughed, saying, "No. I don't want to give up eating meat." The apparatuses of the ontological rift provide all kinds of privileges for many human animals, while leaving a wide and deep wake of traumas. I found her comment understandable and intriguing. Here we have someone who is aware of the traumatic suffering of billions of other species yet turns away from their sufferings so that she can have the privilege of dining on the meat of certain sentient animals in peace. In turning away, in rationalizing, she, like a good Kantian, renders inoperative the categorical command to care for other species. John, by contrast, saw the suffering and trauma Buck was experiencing and felt obliged to stop it and then subsequently to care not only for his wounds but for his overall well-being. Also, John did not stop the violence by traumatizing Hal, but he made clear that violence would result if he did not let Buck go. The categorical command to care means to recognize and respond to the experiences and needs of singular semiotic beings who have their own unique or singular resonant assumptive worlds. In recognizing the singularity of other species, we come face to face with the existential imperative to care, and this requires a great deal of work, work that Kant apparently did not want to do. The work is manifold, psychologically and relationally, yet necessary for resonant kinds of dwelling in the world. The philosopher would be challenged to give up her privileges of superiority and domination embedded in the traumatic exploitations represented in the butcher store or the deli section of the grocery store. It is a change not just in how we perceive other species but in how we treat them. John's care for Buck required a great deal of work and patience.

There is also the work of recognizing our participation in the traumas of other species—in the shattering of their assumptive worlds. And this remorse can accompany the political work of including the lives and realities of other species in our political and ethical discourses. Including them in the zones of both justice and care makes our lives much more complicated. Can we recognize the varied ways we have contributed to or participated in the traumatic sufferings of other species? What individual and collective actions are required if we are to give up our privileges and to care for other species in our area? Where and how do we draw the line in caring for species? What other-than-human animals do we attend to? What about the land, rivers, lakes, and so on upon which all species depend? And, for psychoanalysts, what does all this mean for psychoanalytic education and therapy (Kassouf, 2017; LaMothe, 2024; Searles, 1960)? For instance, some psychoanalytic institutes have statements regarding antiracism, which is a political statement regarding social justice and the traumatizing presence of the ontological rift for African Americans (and other racialized groups). Might justice for other-than-human species be included on websites or in mission statements? Are we interested in how patients understand and relate to other species?

Of course, I recognize that there are times when we have to kill another animal, either because of the danger it represents or because we need food to survive (Nussbaum, 2022). Yet, I would argue that the slaughter of 77 billion animals a year does not accompany any real discussion about minimizing suffering or "humane" killing, let alone a recognition of the horrific traumas they endure. It also does not entail much if any effort to question whether we even need to eat meat to obtain the protein needed to survive and thrive. Add to this the argued necessity for experimenting on other animals.[19] There is a long and tragic history of experimentation on or instrumental use of human beings, usually the most marginalized. The ethical responses to this are all kinds of regulations and guidelines to prevent the injustices of the past. Can we adopt similar guidelines? Or can we at least raise the question, do we really need to experiment on living beings at all? If yes, are we willing to engage in a vigorous discussion regarding the ethics and care for these animals? Regardless, the slaughter of billions of other species, the experimentation on other species, and the absence of ethical discourses are symptoms of the ontological rift between human beings and other species.

To return to London's story, we can see the traumatic consequences of the rift, as well as the possibility of crossing this rift through personalizing, anarchic care. John recognized that Buck (and the other dogs) shared a world together and that Hal's traumatic violence involved privileging his dwelling in the world at the expense of Buck. John's rescuing of and anarchic care for Buck meant that John recognized Buck had a singular assumptive world, which was a basis for his dwelling in the world. The violence of the man who stole Buck and Hal's cruelty shattered Buck's assumptive world. John's care for Buck led to a kind of dwelling that was cooperative and reciprocal, lacking instrumental rationales or relations. The moral of this story in the current climate crisis involves acknowledging the singularity and assumptive worlds of other-than-human species, which invites us to confront the categorical command to care not only for our precarity and dwelling in the world but also for the precarity and dwelling of other species that share and depend on this one home.

Conclusion

Semiosis is a feature of all living beings, and each creature possesses an assumptive world of being-in-itself and being-for-itself (altricial social animals' assumptive worlds include being-for-and-with-others) in relation to other living beings and the Earth. Ideally, this assumptive world comprises a resonant dwelling with others and the environment. Trauma, in general, shatters a living being's assumptive world of resonant dwelling in the world, leaving only bare life or existence. While much of the literature on trauma focuses on human beings—which may itself be a symptom of the ontological rift—I have argued that the shattering of assumptive worlds can be attributed to any living being. Implicit in this argument is the notion that the apparatuses of the ontological rift create a wide wake of trauma, whether we are talking about human beings or other species. The ontological rift unhouses those constructed as "ontologically" inferior, and, in the end, the tragedy of the rift is that it will, if not overcome or rendered inoperative, unhouse us all.

Notes

1 Naturally, there are and were other conscious and unconscious motivations for pursuing psychoanalysis.
2 There is a distinction to be made between suffering and trauma. Suffering is not identical to trauma. That is, not all suffering is traumatic, but trauma is a particular type of suffering. This said, some scholars have explored the question of the suffering and traumas of other-than-human species (e.g., Linzey, 2009, 2013; Scully, 2003; Sollereder, 2020). In this chapter, my focus is on traumatic suffering.
3 There is a philosophical and psychological error in Freud's reflections on his initial and later formulations regarding the sources of trauma. The philosophical error entails equating experience and fantasy as possible sources of trauma. In fantasy nothing happens, unless of course one seeks to realize the fantasy, but then that would move fantasy into the category of dream or vision. Fantasy resides simply and solely in one's imagination and cannot be a source of trauma because one cannot undergo one's own fantasy. Yes, fantasies can be disturbing and troubling, but they cannot be a source of trauma. In an experience, something actually happens. Fantasy is mere cognition, not experience. Thus, fantasy as a source of trauma is not possible. It is analogous to positing that an idea can be a source of trauma. Of course, a traumatic experience is often accompanied by fantasies, usually of rescue, escape, or omnipotent violence, but these fantasies point back to the actual experience of having one's assumptive world shattered.
4 Rank, in my view, seemed to equate psychological disruption, confusion, and pain with trauma. One can agree that birth likely accompanies significant psychological disruption for infants, but it does not necessarily follow that this is traumatic.
5 Will Self (2022) and Parul Sehgal (2022) have recently addressed how the notion of trauma has become a trope in the literary world and the problems associated with its ubiquity.
6 Rosa (2020) has a nuanced view of uncontrollability as it relates to resonance. While the world is uncontrollable, we nevertheless find places and times where we possess a measure of control. In addition, as I previously indicated, the idea of singularity implies uncontrollability. Parents have a great deal of control in relation

to their infants but, when it comes to the singularities of their children, they have no control. If we ever get to the place of completely constructing the genetic features of infants, this will serve as a kind of near ultimate control, which in my view would be monstrous. Rosa, as noted in Chapter 2, indicates that, for resonance to occur, there must be some acceptance of the uncontrollability of the other. In trauma, uncontrollability becomes linked to terror of vulnerability, which means experiences of resonance with the other cannot take place.

7 Another example of communal-cultural trauma is illustrated in Daniel Black's (2015) powerful and disturbing novel *The Coming*. He depicts the horrors of the Middle Passage and the slave markets of the South.

8 One could argue that death represents absolute disintegration, but Winnicott likely thought of death as a state of unintegration. The notion of "disintegration" is, for Winnicott (1945), associated with a failure during the state of primary unintegration (p. 139). He wrote, "I will try to explain why disintegration is frightening, while unintegration is not" (p. 140). In brief, disintegration is associated with trauma and fear or anxiety, while unintegration is not. When Winnicott exclaims, "May I be alive when I die" (as quoted in Kahr, 1996, p. 125), he is, in my view, not associating death with disintegration.

9 These unintegrated experiences are not necessarily the unclaimed experiences that Caruth (1996) associates with trauma.

10 I have changed a number of possible identifying features to protect this person's identity.

11 See Chapter 1, footnote 3.

12 Sigmund Freud and James Putnam had a friendly disagreement about the theory and aims of psychoanalysis. Putnam agreed with Freud that "the physician has no right to impose his own ethical or philosophical opinions on any patient, but must content himself with helping the patient develop in his own way" (Hale, 1971, p. 168). However, he went on to say that "one ethical consideration that seems to me especially significant in psychoanalytic treatment … can be developed without danger, *though not with every patient*" (p. 168, emphasis mine). Putnam believed that raising the unconscious to consciousness had, for some patients, ethical implications for the patient. Freud responded politely, indicating that this perspective would lead away from analysis and the "great ethical element in psychoanalytic work," which is truth (p. 171). Freud was less concerned with ethical aims, perhaps because he wished to avoid religious forms of care and counseling. I think Putnam was mainly correct, which is, in my view, especially the case when we consider the ontological rift and the normative, historical, and political unconscious. To become aware of the traumas of other species means we have to decide what we are going to do. To be sure, we can look away. We can rationalize or justify their sufferings. But we cannot escape responsibility.

13 This is an interesting connection. In the story, violence and domination are yoked to the lawgiver. Philosopher Giorgio Agamben (1998) argues that the law emerges as a result of political violence and the threat of political violence. The "lawgiver" for Buck is the one, representing "civilization," who has the power to dominate and kill without repercussion. In my view, this is an illustration of the ontological rift and its relation to trauma.

14 Philosopher Byung-Chul Han (2018b) writes, "Much more dangerous than the *terror of the other* is the *terror of the same*" (p. 96, emphasis in original). The terror of the same is existential insignificance. These dogs had no singularity—only use value. So, Buck faced physical and psychological trauma associated with violence and being without singularity.

15 Indeed, I argue that London, like Freud, accepted the view that native peoples were uncivilized and closer to "nature." This is evident in his lack of empathy and

understanding of the "native" characters in this novel. London was more imaginatively empathic to Buck's inner life.

16 Interestingly, Fanon (1952/2008) wrote, "The psychoanalysts say that nothing is more traumatizing for the young child than his encounters with what is rational. I would say that for a man whose only weapon is reason there is nothing more neurotic than contact with unreason" (p. 118). In my interpretation, Fanon's claim points to the presence of the ontological rift and its tendency to separate reason from what is deemed to be unreason. In Chapter 1, I noted that Freud's tendency to construct Indigenous peoples as primitive represents, in Fanon's framework, an example of Western subjects' use of reason as a weapon in relation to what is deemed to be unreason. The experiences of Indigenous peoples in relation to Western privileging of rationality have entailed, more often than not, a shattering of their assumptive worlds.

17 I have placed wild in quotes because of the history of juxtaposing civilization and so-called uncivilized peoples (see Canessa & Picq, 2024).

18 London (1903/1981) writes that "in spite of this great love he bore John Thornton, which seemed to bespeak the soft civilizing influence, the strain of the primitive, which the Northland had aroused in him, remained alive and active" (p. 810). In the story, Buck goes back and forth between the civilized world of John (and company) and the wilds or "primitive" areas of Alaska. It may be that London viewed this story as a return to nature, to the primitive, juxtaposing the primitive with the soft civilizing influence. Yet, this interpretation overlooks London's consistent portrayal of the brutality and mendacity of so-called civilized men who are in the grips of the apparatuses that produce the ontological rift. London also does not romanticize "nature" or the "primitive." There is death and suffering as Buck and his pack of wolf-dogs roam the land. But killing other animals is aimed at survival and lacks the motivation of exploitation or wanton cruelty found in civilization. John Thornton is a man of civilization, but he renders the apparatuses of civilization and its apparatuses that produce the rift inoperative through his anarchic care.

19 Experimentation on or killing of animals may appear to stray from London's story, but this is not the case. Buck and countless other dogs were instrumentally exploited for the sake of the privileges and profit of some human beings. Capitalistic rationales were used to justify this exploitation. Similarly, scientific reasoning is used to justify, instrumentally, experimentation on other species.

Chapter 4

Spanning the Divide

Toward Ecological Transferences

In the sixth century BCE, pre-Socratic philosopher Xenophanes "was scathing about Homer and Hesiod because they 'ascribed to the gods all things that are a shame and a disgrace among mortals, stealings, adulteries and deceivings'" (Grayling, 2019, p. 26). Instead of facing our shame and guilt, we transfer it to the gods, excusing or even legitimizing our own foibles and failings. Xenophanes wrote, "If oxen and horses or lions had hands, and could paint with their hands, and produce works of art as men do, horses would paint the forms of the gods like horses, and oxen like oxen" (p. 26). We are not created in the image and likeness of gods, Xenophanes claimed. Rather, we create the gods in our own image. Plato's notion of ideal forms, which nudged Western philosophy down the road of trying to figure out the relation between our material existence and these *fantasized* ideal forms, also appears to fall into Xenophanes's claims of projection. And Plato's Socrates, the Athenian gadfly who sought to challenge citizens' assumptive worlds, apparently challenged political leaders' cherished religious projections. We know the result. Jonathan Lear (1998) contends that "Socrates' mistake … was to ignore the transference" (p. 57) on the part of political leaders who feared public queries of ostensibly sacred truths. I would put it differently. To challenge citizens' assumptive worlds, especially transference-laden religious assumptive worlds, is to invite transference rage and hostility, whether one is a citizen or a leader. Perhaps Socrates did not ignore the "transference" but, rather, simply sought to fulfill his vocation as a gadfly of collective transferences.

Millennia later, Ludwig Feuerbach (1989), like Xenophanes, would take up this idea of projection in his criticism of Christianity. Friederich Nietzsche (Kaufman, 1978) followed suit by identifying the projections inherent to Christianity. Decades afterward, Freud, in his work with Dora (Gay, 1988, pp. 252–253), would use the concept of transference to assess the incompleteness of his clinical work with her. Dora's past experiences, in Freud's interpretation, were "revived, not as belonging to the past, but as applying to the person of the physician at the present moment" (Freud, 1963, p. 138). The patient, in this case Dora, did indeed observe or experience something real about the physician, yet these experiences "will no longer be new impressions, but revised editions" (p. 138). Despite Freud's creative application of this

DOI: 10.4324/9781003518013-5

concept to analytic treatment, projection, displacement, or transferring our fears, desires, and hopes onto other persons and objects appears to be a universal human tendency. In other words, Freud creatively coined and developed the term "transference" in the context of the clinical setting, but the reality of transference in human life had been recognized for millennia by Xenophanes and others.

One of the interesting features about transference is that we are not conscious that we are doing it. We tend to think our experience, our beliefs, and our perceptions are true or accurate representations of reality. For Xenophanes, we, like horses and oxen, think the gods we narrate are real, even if unseen. Our transferences are, more often than not, unquestioned, perhaps unquestionable or, as Adam Phillips (1993, p. 120) writes, a kind of secular idolatry, which, I believe, Socrates was hoping to undermine by raising questions. Of course, we are often more aware of the transferences (specks in the eye) of others than we are of our own (beams in the eye), which only furthers the point about the ubiquitous reality of transference in human life.

To return to Phillips's idea of transference as secular idolatry, transferences include culture and politics and not just an individual's idiosyncratic assumptive world. If cultures, religions, and politics contain innumerable transferences, then our cherished, elaborate philosophical and scientific theories, including psychoanalysis, likewise manifest transference in varying degrees. As Phillips notes,

> With the discovery of transference Freud evolved what could be called a cure of idolatry by idolatry, in fact, potentially, a cure of idolatry through idolatry. But the one thing psychoanalysis cannot cure, when it works, is belief in psychoanalysis.
>
> (p. 121)

Put differently, the "cure" of the patient's transference is effected by the psychoanalytic assumptive world, which is itself transference. Transference, paradoxically, is "cured" by transference.

Yet, the idea of curing is also problematic. "Cure" implies there is an illness, a harm, a cognitive distortion, and so on. Did Athenian citizens need to be cured of their illness? Perhaps Xenophanes and Socrates thought so, but it is doubtful they would have seen Athenian religious beliefs as an illness. More likely, they would have seen their "transferences" as lacking or obscuring wisdom. While there is truth in Phillips's and Freud's clinical perspectives regarding illness, perhaps some transferences are not worth curing or, better, not in need of a remedy. Put another way, maybe some transferences, while exhibiting unconscious material and conscious "illusions,"[1] are actually beneficial, even life-giving and life-sustaining. Can we imagine that kind of transference?

It is important to stress that Phillips's and Freud's views of transference emerged in relation to the context of therapy. Freud (1930/1961b) noted that the development of analytic theory and conceptual tools emerges from clinical work and that there are challenges when concepts are torn "from the sphere in which they have originated and evolved" (p. 144). Freud was referring to the use of psychoanalytic concepts in relation to nonclinical contexts, such as the political sphere. Yet, transference, as noted above, did not originate in clinical work, though clearly Freud and others developed the concept in light of their interactions in the consulting room. The reverse, then, may be true in the sense that concepts that apply to all contexts face challenges when developed in one sphere, such as the consulting room. More specifically, the concept of transference in the clinical context is largely bound by notions such as delusion, illusion, neurosis, psychosis, and reality and truth, which limits the concept. Of course, Freud's (1930/1961b, p. 144) caution about using concepts that originated and evolved in the context of therapy did not stop him and others (e.g., Coles, 1975; Meissner, 1992) from using these concepts as hermeneutical evaluative frameworks for understanding and criticizing culture, history, politics, and so on. However, these varying contexts did not lead to the revision of clinical concepts, such as transference, or psychoanalytic theory.

Transference, in brief, retained its clinical traces when applied to nonclinical contexts, and these contexts did not serve to question or alter the notion of transference. Perhaps we might say that Freud and others *transferred* concepts that evolved in the clinical sphere to other contexts, unconsciously reinforcing the reality and legitimacy of psychoanalysis. Or maybe, ironically, psychoanalytic idolatry can cure, as Phillips suggests, some secular and religious idolatries—idolatries that are symptoms of a kind of transference gone awry. Yet, psychoanalysis cannot "remedy" the ubiquitous reality of transference. More importantly, psychoanalysis needs to recognize and account for those transferences that are necessary for dwelling with other species on this one Earth.

This book is about how an external universal reality, such as the climate polycrisis, can offer a chance to reconsider and perhaps revise psychoanalytic theory and concepts. This chapter focuses on the concept of transference, not simply as it evolved in the clinical context but as it pertains to human interactions with other human beings, other species, and the Earth. Timothy Morton (2017), for instance, provides an example of projections onto other species. He writes, "The myths of the predatory law of the jungle and the alpha male (a now-debunked view of how wolves organize themselves) are Feuerbachian displacements of human ideological capacities onto nonhumans" (p. 172). Morton is referring to human beings' projections or displacements onto other-than-human species and what is deemed to be inanimate matter, which is a focus of this chapter. I will also claim that the concept of transference is a semiotic phenomenon that pertains to other species but also to the polis—a sphere that includes other species and geography or the local environment. More specifically, this chapter reframes *transference as a semiotic phenomenon of dwelling and eco-transference as animistic resonant dwelling*. In addition, this chapter explicates and differentiates between what I call anthropocentric transference and eco-transference.

To make my case, I first provide a definition of transference and depict its attributes, which include the evaluative dimensions that are evident in types of clinical transferences. From here, given the analysis in Chapter 1, I consider psychoanalytic theory and concepts as perpetuating, in part, non-ecological or anthropocentric transferences associated with the ontological rupture between human beings and other species. In other words, psychoanalytic theory functions, *in part*, as an apparatus that produces the normative, historical, and political unconscious that is a symptom of destructive or life-denying transferences associated with the apparatuses that create the ontological divide. This provides the foundation for developing the notion of ecological transference that spans the divide between human beings and other species and that can pertain to any context, including the clinical. Toward this end, I return briefly to a psychosocial developmental perspective as outlined in Chapter 2, arguing that features of eco-transference are evident in the early lives of infants and children, though in the West these are later eclipsed by virtue of the apparatuses of the ontological rift. In brief, ecological or eco-transferences, as I develop them, entail assumptive worlds that (1) maintain the dialectical tension between knowing and not knowing; (2) found non-instrumental epistemologies (or instrumental epistemologies subordinate to personal or animistic epistemologies); (3) foster beliefs of belongingness, multiplicity, cooperation, and reciprocity; (4) produce agencies that exercise positive power; and (5) contribute to resonant experiences of aliveness and a feel of the true in relation to the environment and other species. All of this represents a type of eco-transference that manifests an animistic, resonant life-sustaining and life-giving stance toward other human beings, other species, and the Earth. Ecological transferences, in short, are not in need of a cure—analytic or otherwise.

Some clarifications are necessary before I continue. First, I wish to stress that I am not suggesting that the notion of transference as it pertains to psychoanalytic therapies needs a radical revision. I find the concept, as it is, very useful in clinical contexts. My aim is to emend it, to offer another angle that might alter our perception and use of the concept, clinically and otherwise. Second, it is neither possible nor necessary to offer a history of this concept when others have done so more comprehensively (see Bauer, 1993, 1994; Epstein & Feiner, 1993; Giovacchini, 1993; Tansey & Burke, 1989; Wolstein, 1995a, 1995b). Instead, I engage Hans Loewald's and Jonathan Lear's views of transference, mainly because they connect transference to notions of love and polis. Third, in using the concept of transference, I am also including its twin, countertransference, since they are necessarily the same though they refer to differing roles in the consulting room. Fourth, as I will make clear below, transference involves engaging the world in terms of one's assumptive world, which is a semiotic construction. Since all living beings are semiotic, I contend that transference pertains to all living beings. The clarification here is that to understand the particular transference regarding another species requires understanding the ways they construct and relate to the world.

For a variety of reasons, we are largely ignorant of the psyches, if you will, of other species.[2] The more complex the psyches, the more complex the transference. Since all the attention regarding this concept is aimed at human beings, this chapter focuses on the field of human beings, though, as I claim, it is applicable to all living beings. Finally, aggression, at times, attends transference, assumptive worlds of dwelling, and the exercise of power. When discussing ecological transference, I will spend little time on the relation between aggression and this type of transference, arguing that aggression often emerges when one's assumptive world is frustrated or under threat. This said, as for those transferences that do not reproduce the ontological rift, aggression is used for the sake of a type of dwelling that accepts multiplicity and the potential for reciprocity and cooperation with other species.

Transference and the Polis: Definition and Attributes

Jonathan Lear (1998) notes, "Loewald has argued, transference is human love life as it manifests itself in the social realm" (p. 132). More specifically, Hans Loewald (2000) considered transference a neurosis or illness, as "the patient's love life—the source and crux of his psychic development—as relived in relation to a potentially new love object, the analyst" (p. 311). The "old" love, for Loewald, entails "early pathogenic experiences and their intrapsychic pathological vicissitudes" (p. 308). This does not mean the patient merely repeats these preconscious or unconscious meanings and interactions. Rather, "A meaning is created by the interactions between patient and analyst" (p. 311). The patient, in other words, acts *as if* the analyst is a past love object/person or an aspect of that person. "As if" implies, as Freud made clear in the case of Dora, that there is something real in the patient's recognition of the analyst in the present, yet this recognition is tinged by memories, beliefs, desires, and so on. Put another way, the "as if" of transference is not a mere repetition of the past love but rather a possibility of new meaning that is created in the present in the encounter with the therapist who, in their role and singularity, does not conform to the patient's transference. There are, in other words, interrelated dialectical tensions manifested in transference's *as if*, namely, knowing and not knowing, certainty and uncertainty, invulnerability and vulnerability, the old and the new.

Loewald focused on transference neurosis wherein the dialectical tensions of "as if" are maintained, making possible new experiences. These are the most common types of transference therapists encounter. A seemingly rarer occurrence in the clinical realm is psychotic transference, which represents the collapse of these tensions and the absence of *as if*. The analyst in these transferences *is* the person from the past, and the possibility of new meaning and experience is foreclosed. I will return to this type of transference below because, if connected to Phillips's notion of transference as idolatry, we discover it is quite common, though in more latent and accepted forms in the social world, such as the assumptive world of the ontological rift.

While Loewald framed transference in terms of the patient's love life, he certainly was not overlooking feelings of hatred and hostility directed toward the analyst. Typically, in psychoanalytic literature, these aggressive feelings (and other "negative" feelings and thoughts) are referred to as negative transference interactions, which are part of the patient's love life. Past unrepaired failures or deep disappointments in being loved, in other words, may be evoked in the present. The patient may project or displace onto the therapist experiences of disappointment or failure of the early love object or person. Whether positive or negative, transference represents, in part, resistance as patients attempt to impose their assumptive world on the analyst, trying to force or seduce the analyst to dwell in the patient's world. "The patient," Lear (1998) remarks, "wants Freud to play the role of love; Freud insists on playing the doctor" (p. 71).

Lear (1998) adds to this discussion, arguing that transference "is just the psyche's characteristic activity of creating a meaningful world in which to live" (p. 60). It is not just love or desire but meaning vis-à-vis dwelling that is central here. Since the construction of meaning is a collective enterprise of dwelling in the world, Lear situates the psyche in the polis.[3] "Plato," Lear writes, "thinks that in studying the structure of the polis, we can discover the structure of the psyche" (p. 58). Put another way, the polis "reflects our collective psychic activity. … [W]e have created an environment which our psyches can, for better or worse, inhabit" (p. 67). When an individual "participates in shared cultural activities, those activities will tend also to have an idiosyncratic, unconscious meaning for that person" (p. 69). Lear then contends that "the fundamental demand of all transferences is to participate in the idiopolis, and thus lend it reality" (p. 71). The patient's "as if" is the attempt to have the analyst join in this idiopolis, which would confirm the significance and reality or truth of one's idiopolis—an idiopolis where we dwell together.

Understandably, Lear (and Loewald), as a psychoanalyst addressing other therapists, focuses on transference in light of the analytic relationship. Yet, since he introduces Plato and the polis, we can surmise that transference is situated within the context of the polis (we are political animals, as Aristotle claimed) and that transference can also be collective. That is, a particular polis will construct its own idiosyncratic meaningful world in which to inhabit or dwell, and this engagement with the world can be understood as transference. But this raises questions about the target(s) of the transference. To return to Plato and Aristotle, the "barbarian"[4] is excluded from the polis, and, if barbarians inhabit the polis, then they are excluded from politics. One could say that the Athenian polis involved transferring onto the barbarian those features of the collective that were inimical to Athenian ego-ideals. The "barbarian," like the terms "savage" and "primitive," functioned and functions to confirm the meanings and values of the Greek patriarchal polis (see Canessa & Picq, 2024), which is illustrated in Freud's views of Indigenous (primitive) peoples. That is, collective transference functions to validate the "reality" of the polis's worldview. "Barbarians" were, for the early Greeks, a fact and not a creation, and they functioned to establish the "fact" of civilized Athenians. There is no "as if" aspect of Athenian beliefs about barbarians. This political transference is aimed at exclusion by reinforcing the boundaries and identities of the polis's citizens. If the polis's transference exists in relation to excluded others, does it also occur *within* the polis?

Since I mentioned the Greek polis, let me return to Socrates and his incessant queries, which apparently threatened the secure and seemingly unquestionable meanings of the collective. He was, in my view, attempting to highlight and heighten the dialectical tension between knowing and not knowing, which is necessary for learning and for, in his case, wisdom. Put another way, the citizens, especially male citizens in power, wanted everyone to play the role of the docile citizen ensconced in the rubrics and meanings of the Athenian polis; Socrates, however, insisted on playing the gadfly who sought to test the "as if" nature of Athenian meanings. The Athenian leaders, who apparently feared Aeschylus's maxim "to learn is to suffer," transformed their anxiety and fear into hostility toward Socrates, thereby constructing him as a danger to the polis. Socrates, in other words, posed a danger to the dominant Athenian way of dwelling in the world. There was no "as if" aspect of their political transference and, therefore, no possibility of learning or new experiences. It was, to use Phillips's term, an idolatrous Athenian transference. The execution of Socrates was also a public signifier or warning: in Athenian public discourses there were to be no "as if" considerations about Athenian political-religious meanings and beliefs. Socrates's death ensured the "reality" and continuance of Athenian collective assumptive worlds; his death was likewise the death of wisdom.

There were also less apparent or more occult transferences within the Athenian polis. Athens, like much of the Greek world, was patriarchal. Men were the arbiters or judges of Socrates's fate. Women, girls, and slaves, while included as members of the polis, were constructed as possessing less reason, and this was cited as a reason for excluding them from political deliberations (Aristotle, 1971). It was not "as if" women possessed less reason; this was considered a fact, which, not surprisingly, confirmed the "truth" of men's superiority and their attendant privileges—males' assumptive worlds. While Socrates questioned almost everything else, he never questioned this, in part because it was normative—a normative political transference. Here we see a political transference that was largely overlooked by men and, I suspect, by many women. Patriarchal polities, then, construct apparatuses that entail transferences wherein women and girls are to play subordinate roles while men and boys insist on playing dominant roles. Any cursory reading of history or of present patriarchal societies reveals what happens when women refuse to accept this transference relation.

Transferences within the body politic are ubiquitous and, more often than not, accompany types of unconscious processing—political, normative, and historical. I will say more about transference and unconscious systems, but for now let me note that transference in the polis results in establishing what is normative politically and historically, accompanying a sense of its truth. This inevitably leads to nonnormative individuals and groups within the polis being misrecognized or not recognized at all (as in the case of other species constructed as inferior). The appellation "barbarian," for instance, annihilates recognition of a particular person qua person. They are not included in Athenian history except as anonymous props for confirming Athenian beliefs in their "reality." Similarly, Athenian men are deemed normative, while women are believed to fall short of the norm and are thus subordinate. In the polis, women are included and excluded others, seen occasionally in history. Socrates's story reveals that the polis's secular and religious transferences are highly resistant to being queried.

I am going to return to the notion of transference in the context of political realities, but for now it is important to identify and discuss some of the attributes of Loewald's and Lear's depictions of transference. Whether framed as love or meaning, transference has to do with knowing, which includes beliefs, values, expectations, emotions, and desires. All transference entails some form of knowing; whether this knowing is adaptive, effective, pragmatic, true, and so on, is a question that I address further below. This knowing, which falls under the heading of complex semiosis, is our assumptive world of dwelling, whether we label that an idiopolis, as Lear (1998) does, or the collective assumptive world of the polis in which we live. Since we are finite creatures living in a precarious, ever-changing world, our epistemologies exist in relation to the dialectical tensions between knowing and not knowing, certainty and uncertainty, invulnerability and vulnerability, and the old and the new. These dialectical tensions represent the "as if" aspect of transference. We act "as if" our assumptive knowledge of the world is accurate, useful, protective, and true. The twin evils of Scylla and Charybdis exist at these two poles. If we move toward one or the other pole, the "as if" aspect of our assumptive world vanishes. For instance, the Athenian leaders knew for certain that Socrates was guilty of corrupting the youth and challenging the gods. They knew with certainty that women possessed inferior reason. Their certainty, in my view, was a stab at defending against not only existential uncertainty but also vulnerability. The "old" assumptive world, which Socrates at least raised questions about, must not give in to the new, for the "new" will lead to vulnerability and suffering. The Athenian leaders' transference was an idolatry—an assumptive world or dogmatic knowing that could not be queried. At the opposite pole, there is too much uncertainty, vulnerability, and not knowing, which makes it exceedingly difficult to possess a playful and critical inquiry into one's beliefs and meanings or to exercise or enact with some confidence one's assumptive world.

Lear (2006) provides an example in his discussion of the Crow leader Plenty Coups. Settler colonialism had deprived the Crow people of their lands and their rituals and undermined the narratives that had provided meaning, purpose, and values in the ordering of their lives with one another and in relation to other species and the land. Their assumptive world was no longer relevant given the dominance of the assumptive world of Eurocentric colonization. Plenty Coups, as leader, had to lead his people into a future where there was no clarity about what assumptive world would fit their new world. The Crow subjects' senses of being-in-itself, being-for-itself, and being-for-and-with-others were undermined by the violence of the colonizers. Analogously, in the previous chapter, I mentioned Primo Levi and Jean Amery and their respective traumas at the hands of Nazis. Both men's assumptive worlds were shattered by their experiences, which led to their leading a bare life wherein the "as if" aspect of their assumptive world was replaced by the abyss of trauma.

Implicit in these examples are narratives and rituals that comprise persons' assumptive worlds. Narratives, individual and collective, contain forms of knowing regarding who we are, what we believe, our hopes or aims, and our values (Polkinghorne, 1988). Of course, not all knowledge is in the form of narratives. Mathematics and physics are examples, but even these are situated within larger social-political narratives and their attendant ethoses. For instance, a mountain of scientific knowledge proves that some human beings are the cause of the climate crisis, but this knowledge counters the narratives and associated apparatuses (and their transferences) of imperialistic nation-states, autocratic leaders, and market societies. This cultural-political resistance is explored by a number of analysts (Hoggett, 2010, 2012, 2023; Kassouf, 2022; LaMothe, 2021; Weintrobe, 2010, 2013, 2021, 2024), and I will address this below when I discuss the anthropocentric transference associated with the ontological rift.

All forms of knowing, all semiosis, and all transferences contain not just meanings but also valuations. Loewald's (2000) notion of transference love manifests a positive valuation about some feature(s) of the therapist, which is also joined to unconscious or unstated negative appraisals regarding past experiences or future anticipations of failure. The juxtaposition of differing meanings and valuations is why transference in relation to human beings is complex, layered, and often contradictory. A patient may positively value the therapist, which can screen negative valuations. Outside the therapist's office, multiple, often contradictory, valuations permeate the narratives, practices, and rituals of a community and the larger society. Political transferences manifested in racism, sexism, and classism reveal a conglomeration of valuations based on sets of illusory premises. Racist meanings and valuations designate "whiteness" as a positive, normative cultural valuation based on the illusions that white people are superior and black persons inferior. These are illusory valuations and beliefs because there is no existential basis in reality for them, which is why apparatuses of racism must be created and continually exercised to produce and maintain racist (sexist, classist, etc.) transferences.

In one sense, transference's "as if" represents a stance and method of knowing in relation to others and the environment. Our assumptive world is a semiotic, valuative construction that develops in relation to our immediate environment, and we test out this assumptive world of meanings and values in new situations. That is, we bring our assumptive world to bear in dealing with others, confronting differing contexts, and so on. We act "as if" our assumptive world is applicable in this or that situation. If for one reason or another our worldview does not fit, then we are confronted with emending our assumptive beliefs and values or forcing, by way of aggression, our assumptive world onto others, as evidenced in racist, sexist, and classist transferences. In the clinical setting, the patient acts "as if" the analyst reminds them of their mother, wanting the therapist to play the role of parent. Yet, the therapist, in playing the role of therapist, disrupts the assumptive world (transference) of the patient, eventually inviting the patient to relate to the analyst as they are and not in terms of a wished-for mother.

The "as if" feature of transference, as a method of knowing, is crucial if we are to learn and emend our beliefs about ourselves and the world so that we can survive and thrive. By contrast, Phillips's use of the term "idolatry" in relation to transference represents a stance toward the other (s) that is not one of being open to learning. In other words, the dialectical tension between knowing and not knowing, in this case, collapses toward certain knowing (e.g., Truth). In cultural circumstances where there is an idolatrous transference, a lack of openness and learning results, and the method of knowing is joined to apparatuses of coercion and violence. Margaret Thatcher's campaign slogan, "There Is No Alternative," is an example of this. Neoliberal capitalism (an assumptive world) became the new dogma, or what Harvey Cox (2016) calls the market god. Thomas Frank (2000) makes a similar observation in his examination of "extreme" capitalism by arguing it operates like a religion. As Tony Benn comments:

> If I look at the world today it seems to me that the most powerful religion of all—much more powerful than Christianity, Judaism, Islam, and so on—is the people who worship money. ... The banks are bigger than cathedrals, the headquarters of the multinational companies are bigger than mosques or the synagogues. Every hour on the news we have business news—every hour—it is a sort of hymn to capitalism.
>
> (as quoted in Carrette & King, 2005, p. 23)

The climate crisis reveals how this idolatrous, violent, assumptive world of capitalism, valuing the meaning of capital above all else, is undermining not simply our well-being but also the assumptive worlds and well-being of millions of species.

Transference as a method of knowing in assumptive worlds and other contexts necessarily implies agency. Agency is involved in creating and applying one's assumptive world. Transference love or meaning requires the exercise of one's agency, but this gets increasingly complex when we consider transference and the polis. In the polis everyone has agency, but not everyone has equal power to create or maintain the assumptive world and the attendant constructed apparatuses used to preserve or advance the assumptive world. In the case of market society transferences, the assumptive world of neoliberal capitalism was created and is enacted and enforced by political and economic elites (Piketty, 2014, 2020; Woods, 2017). Factory workers participate in and may accept the assumptive world of capitalism and a market society, but they have little agency to maintain or advance it. If a factory worker leaves the factory floor for work as a labor activist, they exercise their agency by rendering the assumptive world of the factory inoperative, but the assumptive world of capitalism continues to have its negative effects in relation to workers and the environment.

This last point is important because agency suggests accountability for enacting one's assumptive world. In clinical settings, patients (and therapists) are accountable for their transference love and meanings, even if they deny, rationalize, or ignore their contributions. However, at more macro levels, transference and responsibility are more complex. I will say more about transference, the climate crisis, and the ontological rift below, but, for now, not everyone shares equal responsibility for enacting an assumptive world that is implicated in the climate crisis, namely, the assumptive worlds of neoliberal capitalism associated with market societies and the assumptive worlds associated with nationalism's and imperialism's sovereignty. The more agency and power individuals possess within the political-economic system, the greater their culpability. A poor person, living rough, has little agency and accountability for maintaining these larger systemic apparatuses and, therefore, little agency and accountability for climate change. I add here that transferences in early childhood would also manifest less agency and culpability. A newborn infant's assumptive world, which is pre-representational semiosis prior to birth (e.g., a parent's voice), entails a nascent agency, though the child does not yet have the capacity to reflect on and take accountability for their assumptive world and agency.

A further feature and complication of transference in relation to assumptive worlds is conscious–unconscious systems. Freud used the metaphor of an iceberg to refer to these interrelated psychic systems. Consciousness is the tip, and what lies beneath the surface are numerous unconscious beliefs, values, feelings, and memories that shape our present perceptions and behaviors. So, the knowledge and agency involved in transferring our assumptive worlds onto a present context comprise conscious and unconscious semiotic systems. In the clinical setting, patients are not aware that their transference love is, for instance, rooted in unconscious memories of unrequited parental love. Cultural and political transferences comprise conscious conceptualizations regarding the world, as well as unconscious fears, anxieties, and so on. Philosopher Iris Murdoch (2001) remarked that human beings "are anxiety-ridden animals. Our minds are continually active, fabricating an anxious, usually self-preoccupied, often falsifying veil which partially conceals the world" (p. 82). Murdoch believed as well that it was fair to ask political philosophers (and, I would add, psychoanalysts) what they fear because the elaborate conscious constructions they concoct are born in the soil of unconscious and unacknowledged anxiety, fear, and desire. Perhaps the conscious assumptive world of (Freud's) psychoanalysis screens not fear of death per se but our unconscious terror of existential insignificance.

We can surmise the presence of unconscious material with regard to transference when there is a collapse of the dialectical tension between knowing and not knowing, certainty and uncertainty, invulnerability and vulnerability, and the old and the new. There is no more powerful transference than when persons believe their assumptive world is *the* truth, *the* reality, whether this is encountered in the clinical setting or in society. Consider scientist Fred Hoyle, who, along with others, posited the steady-state model or theory of the universe in the late 1940s (Halpern, 2021). Theory is a fancy term for story—a story about the beginning of the universe. While this story faded in favor of another theory or story (the Big Bang), Hoyle held on tightly to his theory, claiming it to be a truer model than others. One interpretation is that Hoyle, after numerous achievements and public accolades, was not aware of his fear of insignificance or that his sense of significance was wedded to his theory. We can also take an analytic stance when encountering white supremacists' rigid grip on racist beliefs, which they regard as unquestionable facts. These are core conscious beliefs that anchor individual and group identity and esteem while screening, in my view, unconscious anxiety and fear or terror associated with their existential insecurity and insignificance.

The conscious–unconscious systems of the psyche or the assumptive world of a person is connected to memory. Transference entails conscious and unconscious memories that shape present self–other perceptions and behaviors. A patient's transference accompanies a preconscious wish that the therapist love the patient in ways they were not loved by their parents. The wish is motivated, in part, by preconscious memories of being hurt and disappointed by parental failures. When it comes to groups or the polis, transference becomes more complicated. A traumatic event, such as the Armenian genocide, is remembered by the repeated telling of the event by those who have no direct memory or experience of it. Present-day narratives, though, shape the transference Armenians have toward Turkish people (and vice versa). Add to this, transference and memory, which are not connected to actual losses or traumas experienced by present-day Armenians, are shaped by these stories that retain the memories of past traumas. From a different angle, in Chapter 1 I addressed how Freud used the terms "savages" and "primitives" when referring to Indigenous peoples. Freud, like many Western subjects shaped by the apparatuses of capitalism, imperialism, and Cartesian–Baconian science, transferred onto native peoples the premise of their inferiority (and the superiority of Westerners). Here is a collective (normative, historical, and political) memory of Western superiority and dominion. This transference falls in the category of idolatry because it is unquestioned and it represents the collapse of the dialectical tensions between knowing and not knowing, certainty and uncertainty, invulnerability and vulnerability, and the old and the new. In terms of memory, past Western narratives are narrated and held in the present as if they were normative, though, as for Socrates, this remains outside of consciousness.[5]

Having mentioned collective memory and transference, let me return to the three categories of the unconscious above and suggest that we consider collective transference in terms of the normative, historical, and political unconscious. In previous chapters, I mentioned Lynn Layton and Timothy Zeddies and their discussions regarding the normative and historical unconscious. I have added political unconscious to this list. These are useful in considering transferences. A normative transference entails a collective engagement in the world that is common, normal, or routine and is not debated or questioned. It is the way we do things. It is the way the world is. Normative transference is connected to both historical (collective memory) and political (ways of organizing the polis) dimensions of human life. It is important to stress here that these terms are not evaluative. Evaluation comes later.

Let's return to Plenty Coups and the Crow people. Before they encountered white European American men and women, they had stories, rituals, rules, and expectations about social organization. These were part of their assumptive world, which included transferring onto each other, other species, and the Earth socially constructed meanings, beliefs, and values that assisted in their survival and flourishing in relation to their environment and other species. Their transference was, for them, normative, historical, and political, at least until they were overwhelmed by the transferences of Western persons. To identify possible unconscious features of these Crow transferences would require delving into their ways of life. In other words, just like in a clinical setting, to begin to understand the unconscious features of a patient's transferences we need to listen to their stories and reminiscences. Understanding, however, does not necessarily accompany positive or negative evaluations.

Since the early days of psychoanalysis, understanding of transference accompanied an evaluative dimension. Since patients suffer, psychoanalysts use diagnostic terms, such as positive, negative, neurotic, illusory, psychotic, regressive, wishful, and not realistic. Above, I mentioned Loewald's view that a patient's transference represents their love life, but clearly this is a pathogenic love that is only partially real. The patient, therefore, needs to have raised to consciousness their unconscious wishes, fears, desires, memories, and so on, that distort perception and agency. That is, the idea of positive transference contains a negative evaluation; the patient's positive transference contains illusions and unconscious or preconscious memories. There is, I argue, a paradox in psychoanalytic clinical evaluations that engenders several difficulties. I think Phillips (1993) was pointing to this paradox when he said that "one thing psychoanalysis cannot cure, when it works, is belief in psychoanalysis" (p. 121). As noted earlier, psychoanalysis is itself an assumptive world that purports to understand and evaluate the assumptive worlds of patients and others. One difficulty is evident: who decides what is an illusion and what is real points to power and authority differences between patient and therapist. Consider the authoritative and possibly authoritarian evaluations Freud made regarding Dora or his categorical statements about savages and primitives. Do we know whether our theories represent apparatuses that produce transferences? Do our theoretical assumptive worlds function unwittingly as apparatuses of the larger society (see Cushman, 1995)? As noted in Chapter 1, psychoanalytic theory, in part, reproduces the assumptive world of the ontological rift, which I will say more about in the next section.

There is another difficulty with psychoanalytic views of transference. I indicated above that, while Freud coined and developed the term, transference is a ubiquitous phenomenon. That is, Freud developed the concept in the context of the clinical setting, which restricted evaluative terms to therapeutic considerations. When the term is used in relation to extra-clinical realities, the evaluative notions come with it. For instance, the transferences of native peoples are deemed to be illusory. Religious transferences are believed to signify a regressive attempt to recover oceanic feelings or Edenic experiences. I am not suggesting these evaluative claims are necessarily false, but they can elide other evaluative assessments as well as screen transference embedded in the evaluation itself. For example, Freud's evaluation of religious beliefs manifested his (counter)transference toward religious subjects (LaMothe, 2004). Let me state clearly that I am not advocating we somehow rid psychoanalysis of evaluative elements, in part because it would be not only impossible but also unhelpful. All human beings, in making use of their assumptive worlds to engage in different contexts, necessarily evaluate, consciously and unconsciously, whether what they have transferred corresponds to or fits with the reality encountered. That is, there is an inherent "reality testing" in transferences. Given this, when assessing transference, I proffer evaluative terms such as "useful," "practical," "resonant," and "helpful" regarding the ways we dwell with each other, other species, and the Earth. These terms are not based on therapeutic evaluations, though I believe they can be used in the clinical context as well. Some clinical transferences may correspond to reality and, as such, may be useful, practical, or helpful. In other words, transferences can be assessed in light of the philosophical question, how are we to survive and flourish in this precarious biodiverse world?

Let me continue with the notion of evaluation. To bridge the clinical and social-political world, *negative transference*, as I am defining it, involves attempts to force and impose one's assumptive world onto others (and other species and the Earth), giving one's assumptive world the air of unquestionable reality or truth. This transference represents, to return to Phillips, idolatry, a collapse of the dialectical tension between knowing and not knowing, certainty and uncertainty, invulnerability and vulnerability, and the old and the new. Racism, sexism, classism, and speciesism are examples of negative transferences manifested in whatever social-political or economic realm they appear. In these examples, there is no possibility of creating something new because vulnerability, in relation to the depersonalized other, is eschewed. Negative transferences result in varied harms to those who are the object of the transference. Clinically, negative transference may be subtle in the sense of attempting to move a patient to adopt psychoanalytic beliefs, thereby undermining and ignoring the assumptive world of the patient (e.g., LaMothe et al., 1998). Positive transference, by contrast, entails using one's assumptive world by accepting multiplicity and engaging others in reciprocal, cooperative relationships, wherein the dialectical tensions (as if) are maintained such that a space is created for correction, as well as the possibility of constructing new shared meanings and so on.

Both negative and positive transference can be framed in terms of utility and correspondence to others and the world, yet I would argue that negative transferences are "useful" (life-sustaining, leading to flourishing) to one group at the expense of others. Positive transferences, generally speaking, are aimed at the mutual well-being of varied inhabitants of the local environment. Also, negative transferences, as I am depicting them, *appear* to correspond to reality, but this correspondence is illusory because it is life-denying to the othered group. The implication here is that "correspondence" to reality refers to social constructions that contribute to life-sustaining and life-flourishing actions toward proximate diverse others, whether these others are human beings or other species. I will elaborate on this further when discussing anthropocentric transference and the ontological rift.

I have been focusing on human beings. Can the concept of transference be used to include other species? In earlier chapters, I accepted the premise that all life is semiotic, which implies that all living beings possess semiotic assumptive worlds that vary in degrees of complexity. Most species' assumptive worlds must be adaptive to their context if they are to survive and thrive—the existential aims of transference and semiosis. This means that transference, for most species, must correspond to environmental realities, and, when it does not, species must change or risk extinction. Unfortunately, the rapid rise of CO_2 and other greenhouse gases since the industrial revolution has led to significant changes in temperature and the degradation of habitats, making it difficult or impossible for species to adapt (Wilson, 2003). Their assumptive worlds struggle to correspond and adapt to a rapidly changing climate and attendant environmental degradations. Human beings are likewise semiotic creatures, though we manifest more complex semiotic capacities, and, thus, our transferences are similarly complex in relation to other species and the Earth. Persons of the ontological rift, for instance, have been successful in forcing (negative power) the environment and other species to adapt to our transference constructions, though, in the end, this negative-anthropocentric transference will be shown to be dangerous and tragic for human beings and millions of other species facing extinction.

Anthropocentric Transference and the Ontological Rift

In this section, I contend that the apparatuses of the ontological rift, as discussed in Chapter 1, can be understood in terms of the concept of transference, which I call anthropocentric transference. In depicting the attributes of this transference, I am working toward identifying its unconscious features, which, in turn, partially explain its resistance to new meanings and new ways of dwelling in the world. In addition, I consider anthropocentric transferences negative transferences because they (1) are idolatrous, forcing, often violently, beliefs regarding the inferiority of other species; (2) lead to a collapse of the dialectical tensions between knowing and not knowing, certainty and uncertainty, and so on; (3) inhibit interspecies cooperation and, therefore, undermine the survival and flourishing of other species (and othered human beings); and (4) tragically undermine the well-being of innumerable species, including humans, by way of degradation of the Earth's biodiversity.

The Western apparatuses of sovereignty produce a collective assumptive world that manifests an ontological rupture between human beings and other species. As noted in Chapter 1, this assumptive world of sovereignty comprises (1) beliefs in the superiority of human beings and the inferiority of other species, (2) the dominance of instrumental epistemologies in relation to inferior others (denial of personal epistemology), and, correspondingly, (3) negative use of power (force and violence linked to domination and subjugation) and agency to enforce this worldview at the expense of "inferior" others. The result of these apparatuses is a severing of human from other-than-human species (and othered human beings). For those of us of the rift, our knowledge of the world and ourselves is embedded in and emerges from these apparatuses. This knowledge (instrumental epistemologies of ruling) and its valuations of inferiority and superiority inform our perceptions, motivations, behaviors, and agency. Because human beings are powerful in constructing and implementing apparatuses, we obtain, by way of the apparatuses of sovereignty and the rift, the belief that our constructions are real or true and, therefore, correspond to reality or nature. Freud's view of "nature," for instance, is an abstraction that Freud believed to be a fact rather than a fictional construction, and it functioned to legitimate the rift. For millennia, then, beliefs associated with the rift were (and continue to be) the "reality" that many Indigenous peoples have experienced and suffered. Colonizers, in other words, have transferred onto colonized persons the assumptive imperialistic world, sometimes resulting in the annihilation of Indigenous cultures and other species.

Let me put this another way. To recall Xenophanes, Western philosophers and theologians created and then projected onto the gods varied conscious and unconscious beliefs, including the notion of nature. That is, Western thinkers have projected onto "nature," Indigenous peoples, and other species meanings and values that have supported instrumental knowing and relations that privilege Western white human beings—beings deemed to be normative, exceptional, and superior. As noted in Chapter 1, the underlying premises of these transference projections are that Western (mainly Christian) humans are (1) the center of the universe and (2) the apex of the created order. This is, in brief, anthropocentric transference.

This anthropocentric transference can be further depicted in terms of the apparatuses of sovereignty and the exercise of negative power. The idea of sovereignty as existentially necessary for the ordering of social life emerged around the same time as the notion of civilization, perhaps 7,000–10,000 years ago. The idea of sovereignty was inextricably linked to Western philosophy and theology or the sovereignty of god(s), which, in my view, was a projection and functioned to legitimate the sovereign and sovereign classes as unquestionably necessary for the functioning of society. The idea of a human and divine sovereign functioned to secure collective personhood and identity within the context of a political-geographical boundary. So, for example, those outside the polis might be human beings but were not Athenian persons. In terms of the sovereignty of god(s), personhood is further abstracted and restricted to a group of human beings. Naturally, the Abrahamic religious traditions, at least in thought but not in practice, construct God (for Christians, a trinity of persons) as founding the personhood of all human beings—at least in theory. The result of this projection was to remove the idea of personhood from applying to other species—species already exiled from the polis except when they had use value. Along with the idea that sovereignty was (and still is) normative and secures human personhood, the idea that othered human beings and other species are persons becomes unthinkable, normatively so. If thought, the idea of other species qua persons is, more often than not, denounced as paganistic, absurd, pathological, or infantile. Consider Freud's view of animism, which he infantilized (declaring it an early stage of development), pathologized (labeling it neurotic in adults), and deemed as something uncivilized primitive people do. Sovereignty, in short, has functioned to secure personhood to human beings (often a group), adding to the severing of human animals from other animals. This, in my view, is a key feature of anthropocentric transference.

A person of the rift might raise the question, are other species really persons? Is this not transference? In Chapter 2, I wrote that the idea of person refers to a unique, valued, inviolable, responsive subject. For philosopher John Macmurray, a person is both a matter of fact and a matter of intention. Parents, for example, in recognizing their infants as persons, intend to treat them as persons (e.g., by caring attunements) and experience children qua persons as fact. In psychoanalytic parlance, parents project or transfer onto their children the idea of personhood. As for other species, personhood would mean recognizing them not as human persons but instead as unique, valued, inviolable, and responsive subjects in their own right. A heron is a heron-person. A mouse is a mouse-person. If this seems a bit far-fetched, we can simply invite people to consider projecting onto a specific living being the notion of singularity and see if it changes their perceptions and behaviors. I will say more about this below.

So far, I have argued that anthropocentric transferences emerge in relation to the apparatuses of sovereignty and beliefs in superiority, which has produced an ontological rift between human beings and other species, excluding them from personhood and the polis or, if included, instrumentally objectifying them in often violent ways—acts of negative power. This raises a question regarding what is unconscious in this transference. Let me approach this first by way of what this transference offers its users. I have indicated that all semiosis takes place against existential insignificance. As human beings, we assign value to each other (ideally) and, consequently, experience being valued. Parents' caring, personal recognitions of their children provide conscious and unconscious organizations of self-respect, self-confidence, and self-esteem (broadly, being-in-itself and being-for-itself). In the West, children eventually internalize the apparatuses of the rift, assigning personhood exclusively to human beings and not other species. As they do so, as mentioned earlier, the normative, historical, and political unconscious develops vis-à-vis other species. But that is only part of the story. Subjects of the rift forcefully project onto other species existential insignificance while retaining the belief in human exceptionalism or ontological significance. To be sure, some species may retain instrumental significance, but they are not believed to possess either existential or ontological significance. Moreover, their instrumental significance is transient—tied to their fulfilling a transient use value. An illustration of this is the denial that other-than-human species have souls, which is a view held shared by many, but not all, Abrahamic religions and some Western philosophies.[6] In short, the apparatuses of sovereignty secure a sense of the inestimable value of human beings, projecting onto other species existential insignificance.

Of course, most human beings, like other species, fear death, but I contend that a deeper fear and anxiety lurk at the very foundation of our Western psyches and anthropocentric transferences, and that is existential insignificance. From birth, if we are fortunate enough, countless acts of personal recognition and care establish our sense of being significant. Our sense of significance is an experience that emerges in relation to the reality of existential insignificance. We know we are finite creatures and will, like other species, die, but many of us face death by establishing a legacy (e.g., children, disciples, books, art), hoping we will be remembered. That is, in facing death we hope we will somehow retain our significance. Yet, death represents the cessation of significance, except for those who remain alive to remember the cherished dead. For instance, in my view, Freud faced death by creating a legacy that would establish his significance beyond his death, but even this legacy is impermanent. A legacy, in other words, has significance in the present and near future for the person who is creating it. When dead, one cannot experience significance, though others who are living may continue to value the person and their legacy. We may "know" this, but cultivating a legacy or hoping to be remembered by loved ones softens the blow of the reality of existential insignificance.

It is not just individuals who face and manage existential insignificance. It is safe to say that human beings create stories, myths, and legends not simply to pass along knowledge and the group's ethos but also to secure a collective sense of significance in the face of existential insignificance. Generally speaking, Indigenous people possess innumerable stories, but these stories are not products of or repetitions of the ontological rift. They do not, in other words, project existential insignificance onto other species or other human beings. To return to Plenty Coups, the Crow people struggled to maintain a sense of their significance when their stories and rituals were stripped away by white settlers who constructed the Crow people as insignificant. It is not just death that troubled the Crow people but being treated as ontologically insignificant. White European Americans, subjects of the rift, projected existential insignificance onto the Crow people, which screened their anxiety and fear of their own insignificance.

The Crow people were not the only targets of settlers' negative manifestations of power or negative transference.[7] Before settler colonialism, it is estimated that the buffalo numbered in the millions. With the arrival of white settlers, the buffalo were slaughtered to near extinction. Most of this slaughter had nothing to do with providing food for people or hides for clothing and blankets. It was, rather, the desire to hunt and kill for pleasure, which represented a clear projection of insignificance onto the buffalo (and other species, such as beavers). This negative transference projection, I am claiming, emerges out of the apparatuses of the ontological rift and sovereignty and manifests the latent or unconscious fear, perhaps terror, of the reality of existential insignificance. It was perceived (unconsciously) to be better to forcefully evacuate this and locate it in othered persons, such as the Crow people, or animals. In this process of evacuation, displacement, or transference, the individual's (or group's) significance, power, and agency were confirmed. Here, note the parallel relation to the transferences of racism, sexism, and classism.

However, all of this reveals that Western personhood, dependent on the apparatuses of the ontological rift, is fundamentally and perennially insecure for two reasons. First, like racism and sexism, the apparatuses of the rift and speciesism must be continually produced to shore up the fabrications of human superiority and the inferiority of othered human beings and othered species. Put another way, the apparatuses of the rift must be continually exercised to strengthen and "normalize" individual and collective "ontological" significance at the expense of those deemed to be without value or of mere use value. Second and relatedly, identity or significance based on the rift is insecure because individuals do not face the reality of their existential insignificance, which is screened by their beliefs in their own specialness or exceptionalism.

The rise of nihilism in the 19th century is another manifestation of apparatuses of the rift's anthropocentric transference and its relation to unconscious fears and anxieties associated with existential insignificance. This rise provides evidence of the problematic relation between existential insignificance and anthropocentric transference. The ascent of science and the decline of Western Christianity as a grand narrative ordering society evoked a crisis of meaning and significance for many people in the West, perhaps best noted in the novels of Fyodor Dostoevsky and the work of Friedrich Nietzsche (see Nishitani, 1982). Similarly, in the early decades of the 20th century, Max Weber (1992) recognized that modernity accompanied the emergence of disenchantment, which he argued is caused by the pervasiveness of a capitalistic society, rationalization, and bureaucratic organizations. Decades later, Jean-François Lyotard (1999) likewise addressed the rise of science, the eclipse of grand narratives, and the resultant mourning in the West. Nihilism was one response to the crisis of meaning and value, which was and is often screened by the quiet desperation of consumerism (Sung, 2007)—as if consumerism can anchor meaning and value.

Offering to explain the roots of nihilism, Nietzsche (1968) argued that the Western human being has "lost faith in his own value when no infinitely valuable whole works through him" (p. 12). His claim about the death of God can be understood as the loss of a comprehensive narrative that had secured a sense of individual and collective personhood and significance. Indeed, the Abrahamic religions have not only ensured (and still ensure) human significance, they have ontologized it, making it permanent (e.g., kingdom of God, eternity, heaven). For Nietzsche, one response to this loss of grand theological narratives is nihilism—life is without meaning or significance. When we consider the cultural devastation of the Crow people (Lear, 2006), we might wonder why their response was not one of nihilism, whereas the loss of Western grand narratives led to the rise of nihilism (and other responses). To be sure, Crow people experienced sadness, grief, depression, and so on, at the loss of their land and their cultural and spiritual practices, but they did not succumb to nihilism.

I think one answer lies in the apparatuses of the ontological rift and anthropocentric transference. For millennia, Western subjects founded their sense of (ontological) significance on narratives that relegated other species (and othered human beings) to inferiority or no significance at all. When these comprehensive narratives collapsed, Western subjects were faced with becoming aware of their existential insignificance. Clearly, God's death, so to speak, precipitated our having to face not simply our existential aloneness[8] but also the absence of our ontological exceptionalism and anthropocentrism. Nihilism, which denies any meaning or value, would appear to be an understandable response to the death of God. Yet, I would contend that nihilism is an avoidance of accepting existential insignificance. In other words, it is analogous to someone who, upon learning that there is no Santa Claus, angrily denies the reality of all gift-giving. The nihilist denies meaning and value, but the truth is that there is meaning and value in life, even though it is impermanent and takes place in relation to existential insignificance. The presence of nihilism, in other words, is an outcome of questioning the foundations of our ontological significance, which are undergirded by the apparatuses of the rift. Moreover, nihilism means that anthropocentric transference (and the attendant rift) is not mourned or overcome, which also means that nothing changes with regard to perceiving the singularities of other species and, therefore, behaving toward them with care and justice. Instead, for the nihilist, the ironic "truth" is that human beings and other species are equally without meaning or value.

I have not yet explained why the Crow did not similarly succumb to nihilism, which is important because it further illuminates the Western roots of anthropocentric transference. I suggest that the Crow, like many other Indigenous peoples, possess narratives that order social life without sovereignty and attendant apparatuses of the rift. In addition, they tend to see other species in terms of personhood or singularity, which means that Crow personhood is not dependent on the existence of a God that exclusively secures their personhood at the expense of other species. In other words, their sense of significance is not solely tied to a belief in God. Instead, Crow persons' individual and collective personal significance is secured to the multiple personhoods of other species and to the land as sacred. Put differently, their grand narratives or origin narratives embrace the multiplicity of singular beings who together face the realities of precarity and impermanence. Singularity emerges from the individual existence of the living being and not from a god or gods. While transference clearly exists in their relation to other species, this is not an anthropocentric, exceptionalist transference, though the possibility of anthropomorphic transferences still exists. That is, their narratives are not attempts to project existential insignificance onto other species while ontologizing their own value. Yes, the cultural devastation associated with settler colonialization undermined their meanings and values, but, as Plenty Coups demonstrated, they just needed some adjustments to retain Crow meanings and values (Lear, 2006).

Might this difference in worldviews between the Crow and Westerners partially account for the enmity settler colonialists had for Indigenous people? In Chapter 1, I mentioned Frank Linderman who, after four decades with the Crow people, felt he could not really understand them. Linderman was the rare white person of his time who cared about the Crow people. The Crow, like other Indigenous peoples, were hated, maligned, sequestered in reservations, and often killed. I think Crow persons (and other Indigenous peoples) represented a threat to the "normative" anthropocentric transferences of white European Americans. Indigenous narratives, in general, embraced a multiplicity of singularities, reciprocity, and cooperation with other species and the Earth (see Deloria, 2003; Erdoes & Ortiz, 1984; Kerven, 2018). They did not differentiate or separate politics and "Nature." Their political philosophies, in other words, were not apparatuses of sovereignty. They held other views of social organization and its relation to other species and the Earth. In my view, subjugating and ridiculing Indigenous peoples—features of the exercise of negative power—were conducted, in part, to provide collective assurance and the normative "truth" or "reality" of Western anthropocentric transferences. In brief, the Crow way of life was understood to be an existential threat to Western ways of dwelling in the world—by exposing the illusions of white European ontological significance. Linderman's difficulty in understanding the Crow people was, in my view, the result of his captivity to the apparatuses of the ontological rift and its attendant anthropocentric transference.

Above, I cited Phillips's view of transference as idolatry, which does not fit all kinds of transferences (e.g., not eco-transference; see below) but clearly fits anthropocentric transference. Idolatry, in psychoanalytic parlance, represents a collapse of the dialectical tensions of "as if," which, in turn, points to profound resistance fueled by unconscious anxiety linked to a crisis of meaning and value. It is not just nihilism that is a response to the loss of comprehensive or grand narratives. Fundamentalisms—secular and religious iterations of belief in epistemological certainty—represent an idolatrous holding-on to anthropocentric transferences and, correspondingly, a resistance to acknowledging existential insignificance. In other words, the use of negative power to enforce assumptive worldviews, such as Western political philosophies, reveals a collective resistance to facing existential insignificance. Nihilism, a modern Western malady born of the crisis of meaning and value, denies meaning and value. In contrast, fundamentalisms ontologically secure the absolute meaning and value of human beings (usually one's own group of human beings) while forcibly denying the singularities of other species and othered human beings. Each response reflects the presence of anthropocentric transference and, concomitantly, an avoidance of the reality of existential insignificance. Idolatrous anthropocentric transference, I contend, is also observed in the pervasive resistance to taking any significant actions with regard to the climate crisis. Of course, many complex factors are involved in this collective resistance (see Weintrobe,

2013, 2021), but I suspect a core feature of this resistance or idolatrous transference is that to do so we would have to radically change our Western assumptive worldviews, which would include facing, in more constructive ways, the cosmic reality of existential insignificance and precarity.

The main point I wish to drive home regarding the relation between existential insignificance, nihilism, and anthropocentric transferences is that there is a fundamental insecurity at the core of Western subjects of the rift. This insecurity can also be understood as dissonant ways of dwelling, of relating to other species and the Earth. Our sense of individual and collective self-efficacy is based on negative power and the illusions of human superiority and domination, which undermine the possibility of reciprocity, cooperation, and attunement to other species and the Earth. Those of us of the rift do not resonate with other species and the Earth. The climate crisis reveals the dissonance of our anthropocentric transferences.

This discussion of anthropocentric transference is not connected, necessarily, to psychoanalytic evaluative terms—terms that were developed in the clinical setting (e.g., neurotic, psychotic). Above, I indicated we can evaluate transferences in terms of whether they are fitting, correspond to reality, or contribute to our flourishing. Of course, anthropocentric transference is felt to correspond to or fit with reality because of the use of negative power to produce this "reality." Moreover, this transference accompanies a sense of being real or true because it leads to the survival and flourishing of many human beings, though at the expense of billions of other species and othered human beings. In one sense, these transferences resonate with us because they accompany a sense of their truth. The climate crisis and the very real near-future possibility of the extinction of human beings, however, reveal that anthropocentric transferences do not, in fact, correspond to the necessary reality of biodiversity nor lead to the flourishing of human beings. Indeed, excluding the flourishing of other species, as a result of the rift, serves as evidence that this transference tragically undermines present and future human persons' survival and flourishing precisely because it does not contribute to maintaining a biodiverse Earth. Put another way, anthropocentric transferences, in the end, dislocate and alienate human beings from ecological dwelling with other species and the Earth. In this sense, we need not use the terms "neurotic" or "psychotic" to recognize that anthropocentric transferences are negative, idolatrous, and tragic.

Before moving to eco-transference, let me briefly contend that anthropocentric transferences are evident in psychoanalytic theories and practice. Recall that Howard Searles (1960) noted the lacuna in psychoanalytic theories regarding the relationship between individuals and the nonhuman environment. Recently, Susan Kassouf (2017) returned to Searles, arguing that psychoanalysts had not accepted Searles's challenge. Susan Bodnar (2024a, 2024b) has taken this task to heart, advocating that we listen to and for the ecological dimension of patients' object-relational worlds, which is, for her, a central clinical task. If we agree that anthropocentric transferences occur in the social-political realm of market societies, then these transferences will be manifested in the clinical setting, oftentimes by the very absence of other species in developmental theories and the corresponding eclipse of analytic curiosity about patients' (and therapists') perceptions about and relations to other species dwelling in the world.

Eco-Transference Spanning the Rift

The term "ecology" comes from the Greek *oîkos* ("house") and λογία ("study of"). Perhaps we could say that ecology means the study of how species dwell in the world. To dwell, for any living being, entails both semiosis and its relations with the environment. The interaction between a living being and the environment comprises transference, acting "as if" semiotic constructions of dwelling correspond to or are resonant with the environment—a correspondence aimed at survival and flourishing. This view suggests that transferences can be understood to be ecological, but I am restricting the term "eco-transference" to the kinds of human transferences that (1) maintain the dialectical tensions of "as if," (2) entail epistemologies that recognize the singularities of other species (animism and belongingness, as well as multiplicity and biodiversity), (3) foster beliefs of cooperation and reciprocity, (4) produce agencies that exercise positive power, and (5) contribute to shared experiences of resonant aliveness and a feel of the true. In brief, I argue that *eco-transference is a semiotic phenomenon of animistic resonant dwelling*. To develop the notion of eco-transference, I return to psychosocial development.

Before infants are born, they have already begun to organize experience semiotically, as evidenced by their preferences for parents' voices discussed in Chapter 2. These early semiotic organizations, I argued, can be depicted in terms of Hartmut Rosa's (2019, 2020) notion of "resonance," which "is constitutive not only of human psychology and sociality, but also of our very corporeality, of ways we interact with the world tactilely, metabolically, emotionally, and cognitively" (2020, p. 31). This "basic mode of vibrant human existence consists not in exerting control over things but in resonating with them" (p. 31). Rosa is not referring to pre-birth or early infancy, but the term is fitting here. Prior to birth, infants' dwelling and semiotically organized preferences are *resonant* with the parental womb. Rosa writes that "resonance in its full sense occurs only when … we feel connected to the world" (p. 33); this connection is understood at this stage as embodied dwelling and belonging. Put another way, early semiotic constructions represent the embodied, pre-representational resonance between the infant and the parental womb.

Once infants are thrown into this buzzing and complex world, they transfer these early semiotic preferences onto parents and the environment with the aims of survival and flourishing. As John Macmurray (1961) mentioned, infants are adapted to being unadapted, which means they are focused on adapting to and cooperating with their parents, upon whom they are dependent. Of course, there are many missteps in the early transferences, whether by infants or parents. These missteps, in which transference fails, can evoke aggression as infants struggle to project their semiotic organizations onto the environment and find a lack of fit, lack of cooperation, or lack of a needed response. Ideally, in these situations, infants (and parents) begin to construct new semiotic organizations that are more adaptable to and appropriate to or resonant with the context.

Implicit in this discussion of early semiotic transferences is the "as if" nature of transferences. To encounter their extraordinarily complex new world, infants use early semiotic constructions to make sense of and engage the world. To survive and flourish, the dialectical tension of knowing and not knowing, uncertainty and certainty, vulnerability and invulnerability, and the old and the new must remain in place if infants are to adapt to the world and to their parents. To be sure, the continuum is more toward not knowing, uncertainty, vulnerability, and the new, but this does not mean infants are without knowledge. Maintaining these dialectal tensions depends on good-enough parents' reliable ministrations, which afford infants a sense of confidence and trust to live in the not-knowing and to be vulnerable. I add that trust and confidence, coupled with maintaining these dialectical tensions, are necessary for ongoing learning to adapt, cooperate, and flourish in a world where we—parents and infants—dwell. By contrast, parental deprivation and impingement undermine trust and confidence and heighten anxiety and relational dissonance, which, in turn, collapse the dialectical tensions. In these unfortunate instances, learning is focused solely on survival.

To continue with this initial period of development, many psychoanalysts accept the view that infants' early epistemology is rife with omnipotence. Winnicott (1971), for instance, argued that infants omnipotently destroy (via aggression and fantasy) the parent or object and that the parent's or object's survival fosters the process of accepting reality and eventually letting go, in part, of omnipotent thinking. Let's linger here and consider whether we need the notion of omnipotent thinking and its relation to aggression in establishing reality or externality.[9] Infants' early transferences entail a nascent agency that already includes some rudimentary recognition of the object or parent as distinct from their nascent self. The point here is that aggression and omnipotent thinking are not necessary for establishing external reality, whether in human infants or other altricial social species. To be sure, we can affirm that infants, as altricial creatures, have an imagination, which is necessary for the ability to adapt to and cooperate with a complex and changing environment. They also have the capacity for aggression. This imagination, though, need not be coupled with the belief in or experience of one's omnipotence. Moreover, imagination is a broader term than fantasy (e.g., omnipotent thinking), which tends to be associated with a binary opposition to reality. These early transference projections, when encountering missteps or failures, evoke aggression and an emending of semiotic (via the capacity for imagination) constructions aimed at resonant cooperation and adaptiveness. In other words, frustration, which is part and parcel of life, means that one's semiotic constructions fail—partially or totally. The agentic response can be one of trying repeatedly to project the semiotic construction aggressively onto the parent or situation or to imaginatively rework one's organization of experience to fit the relational context.

To add to this discussion on this early period of development, I contend that infants imagine the world of objects to be alive as they are alive. Resonance with others and the environment, in other words, depends on this semiotic or presymbolic belief. For lack of a better term, we could call this an early premise that is part of the transference related to dwelling in the world. Infants' semiotic constructions, then, are understood as embodied, relational experiences of resonant aliveness and belongingness. These experiences accompany a sense of the real or the true. For the infant, the world and its objects *are* alive.

While this premise of the aliveness of the world and its objects is an early semiotic construction, it is foundational to later eco-transferences for several reasons. First, the belief that the world of objects is alive suggests an acknowledgment of belongingness amid a multiplicity of objects and contexts (an early form of likeness in difference and difference in likeness; Benjamin, 1995). Second, this premise is evolutionary and ecologically necessary because infants aim to adapt to and cooperate with living parent objects for the sake of their survival and flourishing—dwelling in the world is, by nature, relational, mutual, and cooperative. Third and relatedly, infants, because they are dependent, desire their parent objects to remain alive and, as is evident in infant relationships with depressed parents, will be motivated to enliven the parent to become a parent who represents the aliveness of the world and its inhabitants. In brief, this is an early recognition that one's dwelling in the world depends on the aliveness of the world (as an environmental proximate context) and its inhabitants.

This early semiotic belief is observed in more complex or sophisticated symbolic (narrative, ritual) constructions. George Zachariah (2024) writes that for the Māori "the land is respected as an ancestor, a spiritual being, and an earth mother" (p. 75). Barbara Rossing (2022) points out that the Māori view the Te Awa Tupua River as a person (p. 42). In his work with the Runa peoples, Eduardo Kohn (2013) notes that they believe that other-than-human species "have ontologically unique properties associated with their constitutively semiotic nature" (p. 91). Stoney Nakoda leader Walking Buffalo, quoted by Vine Deloria (2003), asked "Did you know that trees talk? They talk to each other, and they'll talk to you if you listen" (p. 89). Richard Erdoes and Alfonso Ortiz (1984) write, "Just as trees, ponds, clouds, and rocks are thought of as living beings, so the sun, moon, and stars in their firmament are depicted in Indian mythology as alive" (p. 127). Similarly, for many Indigenous peoples, the Earth is a living organism, as well as the foundation (hence "mother") of all living beings. Dana Lloyd (2024) points out that the Yurok believe "the Land itself is a living being. According to Yurok earth-based theology, 'the Earth is alive and has a spirit, and humanity has a vital responsibility to help maintain such harmony'" (p. 121).

These animistic beliefs shape perceptions and behaviors. For instance, Indigenous peoples recognize that their lives depend on a living Earth and its inhabitants, which they believe to be true (as it turns out, it is). Belongingness, multiplicity, and recognition of the singularities of all beings are integral to their psyches and their ways of dwelling in the world. There is *no alienation* between human beings and "nature," *only distinctions.* By contrast, as Vine Deloria Jr. (2003) notes, Western European colonizers have been alienated from nature (and from themselves and others). Dislocation from and dissonance with nature and alienation from other species are part and parcel of how subjects of the rift dwell in the world. Put differently, transferences associated with the ontological rift encounter a world of objects that is mute and dumb. This results in a loss of our self being alive in a world that is alive (Rosa, 2020, p. 22). We lose our sense of resonance because "what appears to us must be known, mastered, conquered, made useful" (p. 6).

Are these Indigenous animistic beliefs iterations of omnipotent or magical thinking? Are they illusions? Are these beliefs remnants of early developmental stages? As for the first question, I would argue that framing these beliefs as omnipotent or magical thinking tacks very close to Freud's views on "primitive" peoples and animism. Moreover, while these beliefs are not illusions (mistaken beliefs), they are not necessarily demonstrable facts, though there is evidence that trees communicate (Schlanger, 2024). If they are not facts, these beliefs nevertheless attend a felt sense of being true or real. The sense of or experience of being true or real emerges when beliefs are assessed in terms of thinking and behaviors that are ecological in the sense of belongingness, multiplicity, and perspectivism, which are joined to shared *aims* of mutual survival and flourishing. We can also consider these beliefs to be pragmatic in that they work ecologically to sustain a habitable environment. By contrast, the beliefs of the rift (superiority/inferiority, dominion) are captive to binary notions of reason–unreason, rationality–irrationality, true–false, real–unreal, and so on. That is, for subjects of the rift, the sense of the true is severed from animistic epistemologies or, if not severed, as in the case of Freud, societally and culturally repressed or dissociated. In addition, I contend that beliefs of the rift are illusions (magical thinking) in that the climate crisis has revealed just how these beliefs undermine the survival and flourishing of human beings and other species. Beliefs associated with the ontological rift are, in the end, entirely impractical since they represent forms of negative power that are leading to the extinction of human beings and millions of other species.[10]

As for the question regarding early development, let me proceed with a developmental perspective regarding eco-transference. In Chapter 2, I mentioned primary transitional objects (PTOs), which are, Winnicott (1971) argued, representations of parent–infant caring interactions. Representations of parent–infant interactions are transferred to PTOs. These objects, like the proverbial blanket, soothe infants during times of anxiety and separation. If Winnicott is correct, the PTO(s) is a living object that is now distinct from but connected to the parent. It is under the infant's control and use, but control does not mean omnipotence. This control functions to provide an emerging sense of freedom and interdependence in relation to an object believed to be alive. I suggest further that the PTO, while not recognized as a person, is recognized as a unique living object amidst a world of living objects. This early type of instrumental thinking is subordinate to an early animistic epistemology that recognizes the uniqueness and *necessity* of this living object in relation to the self. Put another way, the object, which is alive, is believed to be necessary for the infant's well-being, and this well-being is dependent on the well-being of the object owing to mutuality and reciprocity. The infant belongs to the object, and the object belongs to the infant. In this there is an experience of resonance in that they both belong to the environment, which accompanies an embodied, relational sense of the real and the true.

The emergence of the capacities for symbolization (e.g., narratives, rituals) and personalization accompany more sophisticated types of thinking and interactions associated with secondary transitional objects (STOs). These objects are taken from the child's cultural environment, though under their control and agency. The inanimate object is constructed as a living person amidst other living persons—unique, valued, inviolable, responsive subjects. This object, then, represents mutual personal interactions, more expansive resonant interdependence and mutual freedom, and more complex and varied types of mutual cooperation toward survival and flourishing, which includes managing with the STO, by way of imagination, aggression and repair. The early semiotic belief that the world and its multiplicity of objects are alive becomes joined to the capacity to recognize and treat other living beings as persons. Included in this recognition is a nascent acknowledgment of the precarity of all living beings in the face of the reality of existential insignificance and impermanence. Put another way, the STO represents imaginative interactions characterized by a shared exercise of positive power—power and agency aimed at mutual belonging, survival, and flourishing. To return to the comic strip *Calvin and Hobbes* (Watterson & Trudeau, 1985–1995), Hobbes, the stuffed tiger, is a tiger-person (a not-human-person), and the boy and the tiger play, care for each other, argue, and make up in a shared exercise of mutual positive power as they dwell together. And all of this undergirds the nascent sense of mutual precarity, which is countered by mutual care,[11] albeit in the imagination. These mutual, personal interactions give rise to experiences of resonance and an accompanying sense of the true or real in terms of transferences. Interacting with Hobbes, for Calvin (and perhaps for the reader), feels real, feels true, though it is all in Calvin's imagination. The sense of the true and the real, though imaginative, emerges in relation to cooperation, reciprocity, and positive power with regard to dwelling together. This imagination is fraught with transference, which is ecological in the sense of their mutual-belongingness; reciprocal, resonant, and embodied aliveness; and cooperation toward shared survival and flourishing. I add that these imaginative interactions are a necessary bridge for collective cooperative interactions of the group in relation to the environment and its inhabitants.

In Indigenous realms, Calvin's imaginary transferences would easily transition to more complex and varied narratives and rituals that include shared belongingness, epistemologies of multiplicity and perspectivism, and cooperation with a living world of diverse, living, other-than-human persons. In this world, trees talk. Trees, ponds, clouds, and rocks are thought of as living beings, as are the sun, moon, and stars in their firmament. This is a collective ecological transference, and it is true or real in the sense of being environmentally and existentially pragmatic. Put another way, the beliefs associated with ecological transferences, while imaginative, lead to perceptions, relations, and behaviors that are environmentally sustainable for human beings and for other species. Let me stress here that what I call eco-transference includes a shared recognition of the precarity of life and the land (e.g., existential insignificance and impermanence).

By contrast, the non-ecological transferences of Western subjects, once the latter have internalized the apparatuses of the rift, comprise (1) beliefs that they are superior to and have dominion over other species and (2) instrumental epistemologies that secure and privilege human survival and flourishing. To be sure, there is a sense of the real and the true in anthropocentric transferences, but they are tied to illusions of superiority and inferiority, of human dominion, exceptionalism, and exemptionalism. In other words, anthropocentric transferences accompany a sense of what is real and true, but these are connected to underlying dislocation, dissonance, and alienation in relation to the Earth and other species. Capitalism, for instance, is real in the sense that human beings have created apparatuses of market societies that treat species and the Earth as mute and numb, but this sense of reality attends profound alienation. The climate crisis reveals the underlying dislocation, dissonance, and alienation that attend the looming reality of extinction. Calvin, as a Western subject, would eventually internalize the apparatuses of the rift and its instrumental epistemologies, replacing Hobbes and animism with alienation.

I want to tease out two other features of ecological transferences. In Chapter 2, I argued that the STO builds on an earlier epistemology of object knowing and animism, revealing early epistemological principles of multinaturalism, perspectivism, and personalization. These principles of knowing undergird ecological transferences. As anthropologist Eduardo Viveiros de Castro (2017) puts it, "Virtually all peoples of the New World share a conception of the world as composed of a multiplicity of points of view. Every existent is a center of intentionality apprehending other existents according to their respective characteristics and powers" (p. 55). He adds, "Multiplicity can be taken as a kind of plurality [and] Amazonian multinaturalism affirms not so much a variety of natures as the naturalness of variation—variation as nature" (p. 74). These principles lead to more inclusive interactions. These principles, in other words, foster an inclusive polis, wherein spaces of care and justice include other species and the Earth. By contrast, apparatuses of the rift produce anthropocentric transferences that are politically exclusive or restrictive with regard to personalization (and animism), limiting the space of appearances to human beings and usually to a select group of human beings.

Let me offer further comments regarding truth. Above, I indicated that eco-transference provides a sense of the true or real. In general, Western philosophies and sciences have made truth into a logical conceptual abstraction juxtaposed with its binary partner, falsehood. Consider Plato's ideal forms as an early example. What is really true is the ideal form(s). This rational abstraction is divorced from embodied, relational experiences. I add that truth as logical abstraction emerged in conjunction with the rise of the apparatuses of the rift. Consider again Freud's view of animism as developmentally early or a feature of primitive life—as illusory, not true. Freud transferred onto Indigenous peoples the framework that science represents truth and that religion and mythology are illusions. In this transference, Freud, in my estimation, was unable to recognize the illusions present in the ontological rift of the Cartesian–Baconian sciences—sciences that produce what he considered truths or facts. Freud's conception of truth led to an alienation between Western civilized subjects and Indigenous peoples, as represented in his use of the terms "primitives" and "savages." Yes, of course, he argued that all human beings have a primitive side, but he believed that, with progress and science, we leave childhood behind. It is telling that Freud mentioned that adults feel estranged from childhood, as if their childhood is some strange event to be overcome—strange or foreign, like so-called primitive peoples. Similarly, he was unable to grasp the truths of Indigenous stories because of his transferences toward these peoples. Indigenous truths, however, are not simply logical abstractions. Of course, they may be, but they are first and foremost anchored in embodied, relational, and environmental experiences of resonant belonging. In general, Indigenous stories represent ecological transferences—projecting onto other species notions of singularity (also multiplicity), shared aliveness, and inclusive embodied, relational belongingness. The truth of Indigenous narratives is ecologically practical or pragmatic in the sense that these truths help Indigenous peoples to navigate and resonantly dwell in a complex world with the aims of shared survival and flourishing.

The notion of ecological transference may make sense when critiquing Western subjects' nonsustainable perceptions of and behavior toward (i.e., transferences) other species and the Earth. We in the West act "as if" we own the land we inhabit. We construct it as property to do with as we will (e.g., Lloyd, 2024). We act "as if" the Earth is an inanimate or lifeless mute object to be exploited, which exempts us from moral accountabilities (Kant, 1980).

We act as consumers, "as if" the Earth and other species are cheap and without limit. We act "as if" we are superior and other species are inferior, exempting them from zones of justice and politics. We act "as if" we have dominion over what we construct as Nature (e.g., Mill, 1970). Similarly, we act "as if" we are separate from Nature—a term that is a mere abstraction considered to be real. We act "as if" we are the center of the universe or the apex of living beings. All of this has the feeling of being real and true because we have constructed apparatuses that carry out this project. That is, our anthropocentric transferences have produced automobiles, bridges, skyscrapers, flight, space travel, submarines, nuclear power, dams, and so on. All of this seems to confirm the truth of this type of transference. This explains, in part, the social-political resistances toward making substantive changes toward climate action.

Our transferences feel true, but we also resist because we do not want to face squarely the truth of our existential insignificance or the loss of our specialness and privileged position afforded by this transference. Yet, this anthropocentric transference of the rift is devastating ecologically, which the climate crisis repeatedly demonstrates. It produces alienation from other-than-human species and is in the process of leading us to the ultimate dislocation, that is, extinction. Timothy Morton (2017) writes, "Marxism *doesn't work* and therefore *will not survive* without including the nonhuman in its construction" (p. 90). We could replace "Marxism" with psychoanalysis (with qualifications) because psychoanalysis is, in part, a creation of the apparatuses of the West that have produced the ontological rift and anthropocentric transferences. This book is about beginning a conversation toward developing a psychoanalytic theory that moves away from anthropocentric transferences to ecological transferences.

But a question hovers over the utility of these terms in clinical settings, especially given that, in psychoanalytic literature, the notions of anthropocentric transference and ecological transference are absent.[12] Specifically, the term "ecological transference" needs to be developed, and that includes developing its clinical utility. This is a deeply complex question, but I will endeavor to offer a few suggestions. First, it is clear that psychoanalysts need to critique theories with an eye toward whether they unwittingly function as apparatuses of the rift and anthropocentric transferences. Second, while analytically inclined therapists are open to exploring patients' developmental relational experiences, present relationships with friends, families, and partners, sexuality, and religious and political beliefs and values, are we also interested in their beliefs, perceptions, and behaviors toward other species and the Earth? Is there a place for this in psychoanalytic therapies? If so, to what end?

Let me come at this by returning to Freud. In a friendly exchange of correspondence with James Putnam, Freud affirmed that the aim of psychoanalysis is to raise to consciousness what is unconscious. Freud was leery, given Putnam's complaint that psychoanalysis lacks ethics, about imposing something on the patient (Hale, 1971, p. 168). Putnam agreed, but the question for him remained. While I think we should appreciate Freud's sensitivity, we need to acknowledge that, when something comes to consciousness, action, in a sense, is demanded. What is one to do with what one is now aware of? This implies ethics. Moreover, in the discussion above, psychoanalytic theory and therapy are involved in ethics. To be engaged in anthropocentric transferences is an ecologically destructive ethics. It is inherently violent toward other species and the Earth, whether we are conscious of it or not. Psychoanalytic therapies cannot and should not escape ethics, and from here we can invite ourselves and some patients to explore and become more aware of how they perceive and behave toward other species. Once we are aware, then we are faced with a choice to change or not. Will we continue to operate out of an anthropocentric transference or will we find ways toward more ecological transferences that represent *semiotic phenomena of animistic, resonant dwelling*?

Conclusion

For Xenophanes, human beings act "as if" the gods are like us. Freud, by contrast, acted "as if" there is no God or gods. Are these views true? Are our beliefs about God(s) mere reflections of ourselves or illusions, as Freud argued? More apropos than these questions is the recognition that Xenophanes and Freud were pointing to the tendency of humans to engage in transference. We act as if our stories, philosophies, and science are true. Xenophanes had faith in philosophy and critical inquiry. Freud placed his faith in science. He acted "as if" science provided facts or truths and, in privileging science, he demoted religions and animistic thinking to the realm of illusion. In other words, Freud acted "as if" his faith in science and in psychoanalysis as a science was free of transference, but this was itself an illusion, and a powerful one at that, as Phillips (1993) notes. But this is not a critique of Freud or psychoanalysis. All human beings have beliefs, desires, and emotions that shape our perceptions and behaviors. In other words, we transfer our semiotic constructions onto varied contexts, having faith in their truths. Indeed, while this chapter is focused on human beings, I hold that all living beings transfer their semiotic constructions onto the varied contexts they encounter.

To return to human transferences, are we to evaluate our transferences in terms of whether they are fact or fiction? Is Freud's belief in science and rejection of religion true or false? These binaries, a preoccupation of the Western philosophies and sciences that have accompanied the Western colonization of other peoples and lands, keep us, in part, caged and blind. For instance, Freud's transference kept him trapped in Cartesian–Baconian science and prevented him from seeing the truths of animistic epistemologies. Binary evaluative thinking is why Phillips believes transferences are in need of a cure or remedy. This does not mean shedding our interests in discerning what is true or factual and what is not. Instead, we can include and raise to prominence questions about whether our transferences (1) maintain the dialectical tension between knowing and not knowing, certainty and uncertainty, and invulnerability and vulnerability; (2) entail beliefs that lead to recognition of the singularities of other living beings and the multiplicity of nature; and (3) lead to experiences of resonance and cooperation with other human beings, other species, and the Earth. In short, are our transferences (scientistic, religious, cultural, and/or political) ecological, and, if not, how do we make the necessary changes? Put differently, our anthropocentric transferences, which have led to the climate crisis, reveal destructive illusions of the ontological rift and therefore need an ecological correction or remedy.

Notes

1 I place illusions in quotes because I find them problematic, at least in some contexts. Who defines which belief is an illusion? There is a long history of Western colonization of other peoples and their lands. In this sordid history, colonizers often viewed Indigenous assumptive worlds as illusory, while the colonizers believed they were bringing these peoples civilization and reality or truth. Similarly, Freud made claims about the illusions associated with religions, such as the belief in animism. While animism may be an illusion, it is nevertheless a belief that is central to Indigenous forms of dwelling ecologically.

2 There is a great deal of discourse regarding the sentience, consciousness, and agency of other species (Hart, 2024; Monsó, 2024), which for some botanists include plants and trees (Schlanger, 2024). I err on the side that posits some level of awareness that attends the capacity for semiosis, which means that all living beings possess some level of awareness. The more complex a creature, the more complex their capacity for consciousness. Complexity, however, should not be understood in terms of hierarchical valuations wherein the more complex creatures are higher on the taxonomic scale.

3 Lear does not reference Andrew Samuels (1993) who, as mentioned in Chapter 2, has long argued for the inclusion of the political in psychoanalysis. Moreover, there is no mention of Frantz Fanon, who clearly observed the link between the suffering manifested in the consulting room and the political environment characterized by colonial transferences, wherein coercion and violence were used to force the colonized to dwell in the world according to the colonizers.

4 The term "barbarian" is derived from the Greek and was a pejorative appellation for anyone who was not Greek or did not speak the language. The very origin and use of the term are indicative of transference, though having hardly anything to do with love.

5 But didn't Freud say we all have a "primitive" aspect of ourselves? Doesn't that suggest there is no transference? Perhaps that was a step in the right direction, but what is retained are the premises of superiority and inferiority, as well as of dominion. As I explain in Chapter1, Freud considered what he deemed to be primitive (e.g., animism) to be surpassed in human development; if not, Freud pathologized it (e.g., neurotic). Put another way, the apparatuses of Western society were implicated in producing the collective memory, and transference remained unchallenged. This had devastating consequences, which I will elaborate later in this chapter.

6 Aristotle (1987) believed plants possess a vegetative soul, animals a sensitive soul, and human beings a rational soul. His notion of soul is tied to both taxonomic hierarchy and sovereignty, which, he believed, is a feature of animate and inanimate existence (Aristotle, 1971, p. 12). This unified view of nature, nevertheless, possesses apparatuses (hierarchical valuations and sovereignty over other species) that produce the rift between human beings (and othered human beings, such as enslaved persons) and other species.

7 Daniel Black's (2015) novel *The Coming* illustrates the relation between the capitalist imperialism of the slave trade and the struggle of enslaved Africans to survive while also retaining some sense of their significance in the face of terror and humiliation. The hatred of the slavers toward their "cargo" represents their unconscious fear or terror of their own insignificance.

8 Camus (1956/1991, p. 250) argued that our metaphysical rebellion has resulted in our (Westerners') experience of being alone. Indigenous peoples, because of their animistic, personal epistemologies, would be less susceptible to existential aloneness since they view the land and its inhabitants as kin.

9 I think the notion of omnipotent thinking is a relic of the Western ontological rift. It implies absolute power (sovereignty) over the object—a relation of complete domination. It is not clear to me that infants, in fantasy, seek absolute power over the parent. To be sure, infants, like adults, seek some measure of control or agency in pursuing their needs and desires, but this does not mean we need to construct this as omnipotent thinking. I do think there are examples of imagination that include omnipotent thinking in adult life. The entertainment industry is rife with examples, as are forms of slavery, exploitation, etc. Yet, I do not think we have to believe that omnipotent thinking is rooted in early development. Indeed, I contend that omnipotent thinking in adulthood is a kind of imagination that is not ecologically adaptive because it is rooted in desires to have power or dominion over others and other species—overlooking their singularities and needs. Put differently, I believe the apparatuses of the ontological rift are founded in a type of imagination that depends on omnipotent thinking. A final reason for moving away from the idea of omnipotent thinking in early childhood involves other altricial species. Can we reasonably argue that other infant primates, for instance, have the capacity for omnipotent thinking since they are our closest relatives? Is omnipotent thinking necessary for the creation of external reality? Is that its evolutionary role? I think it is better to argue that other infant primates, like human infants, have the capacity for imagination, which is necessary to adapt to complex and varied environmental contexts. That the imagination can go awry, in human beings and other species, does not mean we need to posit omnipotent thinking.

10 I am not suggesting we rid ourselves of sciences or philosophies that struggle with identifying what is true. It is possible to view and conduct science that respects and values the singularities of other species, as well as to acknowledge its limits with regard to claims of the universe or the Earth as living. Let me add that the notions of truth and self-deception remain important for psychotherapy. Therapists and patients, like other human beings, can deceive themselves (Fingarette, 1969/2000) or deny a truth. My point here is that we need to frame truth in relation to our ecological dwellings in the world.

11 This "care" is joined to parent–child caring interactions that are organized and projected onto the STO.

12 In a search of the Psychoanalytic Electronic Publishing library, which has over 151,000 articles and 138 books, there is no mention of ecological transference. A Google Scholar search yields two articles, but they are not psychoanalytic.

Chapter 5

Reimaging Resistance as Acts of Impotentiality

The king of Thebes, in the Oedipal myth, is confronted by the prophet Tiresias regarding his killing of his father, King Laius, and marriage to his mother, Jocasta. Similarly, the prophet Nathan tells a story to King David about a man who steals another man's only sheep. David is outraged, realizing a moment later that the theme of Nathan's story is really about David having his general, Uriah, killed so he could marry Uriah's wife, Bathsheba. In Mary Shelley's (2000) *Frankenstein*, the death of his mother initiates Frankenstein's manic-obsessive attempt to find a way to reanimate life, which leads to the creation of a "monster," the deaths of people he loves, and eventually his own death. Captain Ahab, in Herman Melville's (1851) *Moby-Dick*, is possessed by and captive to his desire to reap vengeance on the white whale, which leads to his death and the deaths of almost all the crew. In Charles Dickens's (2013) *A Christmas Carol*, Ebenezer Scrooge, having experienced the death of his mother, followed by his father sending him away to school, pursues wealth with a dour determination that shields him from past, present, and future losses. In Dostoevsky's (2009) *The Brothers Karamazov*, the Grand Inquisitor interrogates Jesus. The narrator says,

> When the Inquisitor ceased speaking he waited some time for his Prisoner to answer him. His silence weighed down upon him. He saw that the Prisoner had listened intently all the time, looking gently in his face and evidently not wishing to reply. The old man longed for Him to say something, however bitter and terrible. But He suddenly approached the old man in silence and softly kissed him on his bloodless aged lips. That was all His answer. The old man shuddered.
>
> (p. 654)

The Grand Inquisitor, who seemingly holds all the cards, anticipated angry confrontation but not the prisoner's response. Neo, the character in the movie *The Matrix*, is faced with a decision: take the red pill or the blue pill. If he takes the blue pill, he will wake up at home in his bed—in a world of illusions. If he takes the red pill, he will find himself in reality with all its struggles and

DOI: 10.4324/9781003518013-6

uncertainties. Neo swallows the red pill. Numerous others in real life have refused to take the blue pill. Frederick Douglass, Sojourner Truth, Eugene Debs, W. E. B. Du Bois, Ida B. Wells, Fannie Lou Hamer, Rosa Parks, Frantz Fanon, Amilcar Cabral, C. L. R. James, Aimé Césaire, Paul Lafargue, and untold others preferred to face reality and to reject and resist the tyrannical power of racism, imperialism, sexism, and classism.

Whether in myths, novels, movies, or real lives, all of these represent types of resistance in human life. Some resistances, whether individual or collective, are defenses against the pain of guilt, shame, and loss that leads to tragic ends—unless faced and worked through. Other types of resistance are responses to the apparatuses of oppression—apparatuses that are themselves manifestations of destructive resistance. Often, they are intertwined. The point here is that the notion of resistance, like transference, is a feature of human life and not simply or solely applicable to clinical interactions. If resistance is a universal reality, then how do we understand its necessity and functions in life? Is there an evolutionary and ecological value(s) to resistance? Is resistance an existential feature of semiosis? Do other species resist? What is resistance's relation to epistemology or, more broadly, to semiotics as it relates to human beings? What is resistance's relation to human freedom? What if resistance is not only a psychological defense but a necessary feature of creatively engaging the world? These are the questions I am interested in taking up in this chapter. Since the notion of resistance is another key concept in psychoanalysis, I begin by attending to its meanings, attributes, and limitations. From there, I consider resistance in light of the ontological rift, semiosis, and the consequences of resistance, namely, the disquiet dissonance of Western subjects—a form of defensive resistance against the reality of existential insignificance. This provides the foundation for expanding our understanding of resistance by returning to psychosocial development, animistic epistemologies, and the work of Giorgio Agamben. These are key for depicting ecological-political forms of resistance—forms that render the apparatuses of the ontological rift inoperative. All of this provides the basis for the next chapter on the cultivation of ungovernable animistic selves in the Anthropocene Age. I conclude with some brief thoughts regarding psychoanalytic training and practice.

A few words before beginning. Psychoanalysts have long recognized that transference and resistance are closely related. Transference, in other words, is a form of resistance (Frank, 2012; Lentz, 2016; Sherman, 2007). Indeed, George Frank (2012) contends that transference and resistance are foundational to psychoanalytic inquiry, both theoretically and clinically. I have devoted a chapter to this term because exploring and amending this concept can expand our understanding of a psychosocial development that is not captive to the ontological rift. Positively stated, reimagining resistance can move us toward understanding more ecological ways of dwelling in the world. This raises another concern. Does reimagining resistance undermine the utility of this concept for psychoanalytic therapy? I intend to argue that it does not, though it does provide a more expansive view of resistance—one that is applicable to extra-clinical realities and is not simply framed in terms of psychosocial defenses. In other words, as with transference in the previous chapter, this exploration will offer more evaluative terms when resistance is assessed. Let me also mention that, like the notion of transference, resistance can also refer to the therapist in the form of counterresistance (Schoenewolf, 1993). Since it is the same concept, there is no reason to differentiate between the two in this chapter. Lastly, in the Psychoanalytic Electronic Publishing database, which has over 151,000 articles, there are 22,413 mentions of the term resistance. In addition, nearly every psychoanalytic book on the theories and practices of psychoanalysis uses this term (Aron & Mitchell, 1999; Chessick, 1991; Gabbard, 1994; Kernberg, 1984; Langs, 1989; Slipp, 1991; Wolfberg, 1995). I mention this not simply to acknowledge the pervasiveness of this concept but also to point out that it is not possible, nor do I think it is necessary, to provide a comprehensive overview of this concept in the context of psychoanalytic practice and theories. I can, however, provide a brief overview of its attributes and its relation to life and the ontological rift.

Resistance and Its Attributes

In his case histories, Freud (1895/1955d) recognized that patients who consciously sought treatment also resisted the psychoanalytic process, signifying the presence of unconscious fears and anxieties. Some years later, Freud (1912/1958a, 1913/1958b, 1914/1958c) developed the concept of resistance in his papers on psychoanalytic technique. In general, Freud (1900/1953) contended that "whatever interrupts the progress of analytic work is resistance" (p. 517). A slight adjustment to this came later: Freud (1926/1959c) categorized "all forces that oppose the work of recovery as the patient's resistance" (p. 223). Since this covers quite a bit of territory, identifying some of the landmarks will help us navigate our way to a better appreciation of the strengths and limitations of this key concept.

For Freud (1926/1959b), "The analyst has to combat no less than five kinds of resistance, emanating from three directions—the ego, the id and the superego" (p. 160). In terms of the ego, there are three resistances, namely, repression resistance, transference resistance, and gain from illness. The last indicates that people develop psychopathologies to survive, suggesting that these illnesses, while not satisfactory, have an inherent value. The other two resistances derive from the id and the superego, and the superego "is also the most obscure though not always the least powerful" (p. 160). For Freud, identifying the type of resistance was important for understanding the sources that shape interpretations and interventions. Yet, the overall challenge for the patient and the analyst is to work through resistance for the sake of cure, health, and so on. Working through resistance (and transference) is, then, a central feature of the psychoanalytic process.

Resistance has several important aspects. First, for Freud, resistance is an intrapsychic and interpersonal phenomenon (Gill, 1982) that is manifested in the clinical encounter and/or in other social venues. Intrapsychically, the ego can resist the id or superego and vice versa, all the while resisting the therapist. Human beings seem to be capable of a plethora of intrapsychic and interpersonal resistances. Second, resistance is conceptualized primarily as a defense—a defense against whatever is unconsciously deemed by the patient (or the therapist) to be a threat—real or imagined. Third, the analyst, ideally, identifies the resistance and its sources as part of the task of facilitating the process of working through it. Fourth, anything that disrupts or interferes with this process is considered resistance, whether that is a patient's love for the therapist, hatred, compliance, agreement, disagreement, being late, being on time, or fantasies (Panetta, 2002). In other words, almost anything the patient says and does can be interpreted by the therapist as resistance, as long as it falls under the heading of obstructing the process. Of course, who determines whether there is obstruction and what the source of the obstruction to the process is are of great importance. Finally and relatedly, while Freud wrote that "whatever" impedes the process is resistance, the focus was primarily on the patient as the source. Later, therapists were understood, at times, to be a source of resistance in the analytic process through the process of counterresistance.

These last two features point to the interpretative authority of the analyst, which Freud recognized. He believed the authority of the analyst could be leveraged to assist patients in becoming aware of and working through their resistance.[1] This led many early analysts to view the patient's resistance as undermining the authority of the analyst (Adler & Bachant, 1998), which was later critiqued by self psychologists (and others) who believed resistance was, in part, a necessary feature in establishing and maintaining a self. Self psychologist Hans Kohut (1984), for example, wrote, "My personal preference is to speak of the 'defensiveness' of patients—and to think of their defensive attitudes as adaptive and psychologically valuable—and not of their 'resistances'" (p. 114). Kohut and others (e.g., Samberg, 2004), in my view, were trying to move away from Freud's more authoritative, negative appraisal toward interpreting patients' behaviors as resistance, as well as to considering positive aspects of resistance in clinical encounters. Perhaps this shift in hermeneutical authority resulted from more egalitarian, democratic tendencies in the United States as well as decolonial liberative movements after World War II.

There is more to the story. Howard Bacal (1998) later added to Kohut's view. "From the perspective of self-psychology," Bacal wrote, "resistance to the analytic process is seen as reflecting a fear of retraumatization through repetition in the analytic relationship of traumatic childhood experiences" (p. 23). Here we note a movement away from Freud's abstract intrapsychic sources of resistance to an understandable resistance related to the fear of being retraumatized in the analytic process, whether this is a resistance to remembering or, having remembered the traumas, a resistance to the therapist's care. This later view of resistance can be understood in terms of vulnerability. Trauma occurs at the point of vulnerability. Persons who have been traumatized in childhood understandably resist being vulnerable (and uncontrollability) since it is equated with trauma. In therapy, part of the process of working through the traumas requires vulnerability. Time after time, I have seen patients struggle, sometimes with anger and hostility, because they are terrified of vulnerability, even when they have come to trust the therapist. The reality is that healing from the wounds of childhood, as well as genuinely experiencing and integrating the care of the therapist, requires vulnerability. But this vulnerability is resisted because vulnerability is, in the psyche, equated with trauma and uncontrollability. As Helmut Thomä and Horst Kächele (1994) write, "Resistance is related to the change which is consciously desired but unconsciously feared" (p. 99). In brief, for Thomä and Kächele (and Bacal), resistance functions to regulate relationships and protect the self (pp. 103, 105). That is, resistance represents an attempt to foster a desired response or to avoid a frightening behavior from the other (Horner, 2005). While resistance can refer to interfering with the process, it is also viewed by self psychologists as an understandable and, at times, healthy psychic activity (Gabbard, 1994, p. 102).

The change in how resistance was understood led to modifications in how to respond. Freud sought to identify, understand, and work through resistance by using the authority of and trust in the analyst. This is largely held to be the case for many analysts following him, though later therapists considered that there were times to side with the resistance. This approach is based on understanding that some types of resistance need to be not only understood but also appreciated for the positive or constructive psychic-relational functions they perform (Sherman, 2007).

Adam Phillips (1993) adds to the discourse on resistance in his discussion about obstacles in life and their relation to knowing and the self. Phillips writes, "The first relationship is not with objects but with obstacles" (p. 85). In one's first relationship, the "way we get to know what we eventually call a mother is through the obstacles of her presence. … Consciousness is of obstacles" (p. 91). More particularly, Phillips argues that "the child can find out what the object is—or rather, get a version of what it might be, its properties and interrelations—only by finding or constructing obstacles to access or availability" (p. 90). In my reading of Phillips, obstacles by their very nature are necessary for coming to know (and love) objects-persons and for the emergence of the awareness of objects-persons. I will say more about this below when addressing resistance in terms of semiosis and psychosocial development, but for now it is important to stress that resistance is an existential feature of both consciousness and knowledge of self, other, and the environment. The ground beneath us resists our steps, making it possible to walk down the street. Strangers resist my projections, creating a space for me to come to know them. A possible friend, in his singular being (being-in-itself and being-for-itself), resists, and, in his so doing, I come to know him.[2]

Phillips is naturally aware that obstacles are not neutral and resistance is not always defensive. In other words, we need to consider what kind of obstacle we are encountering because "poor obstacles impoverish us" (p. 86). By contrast, a good-enough parent is an obstacle that enriches the child. The parent's reliable ministrations, which are themselves a feature of the parent as obstacle, create a space for the infant to be-in-itself and be-for-itself—to actualize the potentialities of an infant's singularity. Put differently, good-enough parents provide children with a sense of trust that accompanies the unconscious existential belief that obstacles are necessary for knowing and dwelling—for being-in-itself, being-for-itself, and being-for-and-with-others. The reverse is also the case. The parent encounters the singular infant as an obstacle who resists the parent's fantasies and, in so doing, the good-enough parent ideally comes to know and love the child in itself.

Of course, there are obstacles we encounter and obstacles we may choose that are detrimental to our well-being. In terms of psychosocial development, early parental deprivation and impingement mean that the parent is a bad object and also a bad obstacle. By "bad," I mean that children will, if they survive, adapt such that obstacles, including themselves, will be something to be feared. They will be left with a desiccated sense of self and other. Consciousness and knowledge will be confined to survival, thwarting the actualization of other potentialities. I recall a woman who had had an incredibly abusive father and an emotionally absent mother. These experiences left her fearing others, whom she perceived as having power over her. She survived by choosing obstacles (e.g., pets) that she had power over, providing them with the kind of care she longed for but did not receive. During the first two years of therapy, she resisted experiencing these past horrors of deprivation, not simply because these memories were painful but also because I had not yet become a good-enough object or obstacle to witness and work through these memories with her. Indeed, there were times when she viewed me as the bad object or obstacle, which I understood as her getting in touch with her vulnerability—her earlier vulnerability was equated with trauma. Paradoxically, I was an impoverishing (bad) and good-enough object and obstacle.

There are other illustrations from literature about encountering painful obstacles and choosing negative obstacles. The prophecy of Tiresias was an obstacle that King Laius sought to remove by having his son, Oedipus, killed. Captain Ahab obsessively pursued the white whale, in the end killing himself and all but one member of the crew (someone had to survive to tell the story). Ahab's knowledge of himself (and the crew), evident in his maniacal pursuit of the whale, and his knowledge of the whale (a mere enemy to be defeated) were profoundly and tragically impoverished. His consciousness, in other words, was severely restricted by his obsessive pursuit of the white whale. Young Frankenstein encountered the obstacle of grief and turned away, creating what he believed would overcome death and loss. Instead, he created a monstrous obstacle, destructive to those he loved and eventually to himself.

Frankenstein's self-knowledge was diminished by his refusal to face and work through grief.[3] And, when the creature came to life, Frankenstein refused to come to know and care for the creature. Frankenstein was himself an impoverishing obstacle, which the creature eventually tried, but failed, to convince Frankenstein of. The monster's loneliness and malicious envy consumed his sense of self and consciousness, all of which resulted from encountering depriving obstacles, whether Frankenstein or the villagers. Dickens's (2013) Ebenezer Scrooge chose the obstacle of making money to avoid and resist coming to terms with earlier painful obstacles of loss. His diminished knowledge of self and others was evidenced by his lack of empathy toward Bob Cratchit and his family, Scrooge's nephew, and poor people struggling to survive the obstacles of capitalism. The ghosts of Christmas Past, Present, and Future were positive obstacles that helped Scrooge overcome his resistance so he could become conscious and learn about himself and others. The ghosts, in one sense, established a relationship and process whereby Scrooge could face and work through his resistance to past losses. Some obstacles may be painful and frightening, motivating us to resist by way of seeking other obstacles, which, as these stories illustrate, often lead to tragic endings.

The examples above involve individuals. However, resistance—impoverishing or life-enhancing resistance—and obstacles can also be collective. Sexism constructs women and girls as obstacles, both feared and desired. Women are feared because, by their very existence, they expose the falsehoods and self-deceptions of male beliefs in or, more accurately, illusions of, their superiority. Men who believe themselves to be superior resist facing their own precarity and existential insignificance. Their self-knowledge and knowledge of women in their lives are distorted. Male desire is awakened when women as obstacles submit to inferiority, undermining women's sense of self and knowledge. Male hatred and hostility, which are evidence of resisting and repressing fear and anxiety, emerge whenever women stand up for themselves, thus resisting the obstacles of male privilege and power. Colonizers, as Frantz Fanon (1952/2008) knew well, believed themselves to be superior to native peoples. When native people started to resist, they were constructed as rebellious obstacles to be overcome through violence and the threat of violence. For Fanon, the goals of therapy, in his context wherein patients had suffered at the hands of French oppressors, were (1) "to '*consciousnessize*' [the patient's] unconscious, to no longer be tempted by a hallucinatory lactification," and (2) "to enable [the patient] to choose an action with respect to the *real source of the conflict*, i.e., the social structure" (p. 80, emphasis mine). Put another way, Fanon sought to help patients (and peoples) to become aware of and resist the bad obstacles associated with colonialism and to choose more life-giving and life-affirming obstacles associated with political freedom and human flourishing. These life-affirming obstacles would have been seen by colonizers as resistance to the "truths" of the French belief in their own superiority. As Lewis Gordon (2015) writes, "Fanon introduced the importance of experience and systemic resistance … his counsel is, in short, actional" (p. 71).

There is more to unpack regarding Phillip's notion of obstacles and resistance. Phillips states that consciousness arises as a result of encountering obstacles. Broadly speaking, one implication is that to be alive is to encounter resistance and obstacles. Put another way, the very existence of semiosis and the corresponding senses of being-in-itself and being-for-itself necessarily mean there is resistance. By contrast, death is the absence of semiosis and resistance. Similarly, sameness, which is impossible though feared, is the eclipse of resistance. "A total loss of resistance," Byung-Chul Han (2018b) writes, "would level the other to *the same*" (p. 45). In one sense, death and sameness represent neither obstacles nor resistance, though both evoke the anxiety that is a source of resistance. Consider the notions of annihilation and engulfment anxieties. Both, in my estimation, can be understood as the threatened denial or absence of one's senses of being-in-itself and being-for-itself, which, in turn, means the eclipse of semiosis and good-enough resistance. One may object and argue that the source of these anxieties is impinging or depriving parents. These are obstacles. They overwhelmingly resist the child's sense of self. Traumatizing objects/persons represent the kind of obstacles that pose a threat to life, semiosis, and singularity, whether through the threat of death or that of sameness. In other words, the traumatizing obstacle is the object that forcibly reveals the possibility of the eclipse of semiosis—death and sameness. It is the obstacle that is resisted for the sake of safeguarding some semblance of being-in-itself and being-for-itself (and, collectively, being-for-and-with-others).

A few illustrations will be helpful here. Anyone familiar with Star Trek's *Next Generation* series will recall Jean-Luc Picard's encounter with the Borg collective. The Borg assimilate other species into the collective, destroying their individuality or uniqueness. Each member of the Borg exists merely as a cog, retaining value only to the degree that they (there is no gender among the Borgs) serve the survival and other aims of the collective. This may remind us of the Borg-like reality of capitalism and its insatiable assimilation of people and cultures. The Borg evoke horror in human beings, including Picard, who was assimilated and later rescued. It is not simply death that is frightening. Indeed, death, for Picard and other human beings, is a better fate than being assimilated in a form of living death. At least, in choosing death, one is manifesting singularity and self-efficacy. As Han (2018b) writes, "*Much more dangerous than the terror of the other* is the *terror of the same*" (p. 96, emphasis in original). It is futile to resist, the monstrous Borg queen[4] tells Picard and others. The Borg can be considered an obstacle that resists or, better, annihilates the semiosis of singularity. Put differently, to be assimilated means there is no semiotic activity except that of the collective and, hence, no resistance. The individual Borg, in other words, does not construct experience. There are no consciousness and no self-knowledge, as well as no knowledge of the other. The individual Borg is a monstrosity because it is alive without the very thing that constitutes life, which is semiosis, singularity, and sentience. Sameness may be impossible, but that does not mean human beings do not attempt it when seeking to depersonalize other peoples (and species). Colonizers, for instance, were and are like the Borg.

A related illustration is zombie movies. Zombies are the living dead—an oxymoronic horror. Zombies feast on the living, turning many of them into the living dead. As Han (2018a) writes, "Without the negativity of death, life solidifies into something dead" (p. 45). Like the Borg, zombies, or the "living" dead, represent the absence of singularity and semiosis. They are all basically the same, and the same is equated with an absence of semiosis and, hence, of life—but it is bare life. Jon Greenway (2024) argues, in part, that zombies and other monstrosities of literature and film are connected to the necro-politics of capitalism. Capitalism and market economies reify individuals to sameness. Workers have instrumental value to the degree they contribute to profit and market expansion. They are assimilated and are without singularity, making it easier to discard them when they are no longer of use. The categorical imperative does not apply to them, as it does not apply to the Borg or zombies. The Borg and zombies are not due care or justice because they represent the annihilation of singularity. They, like capitalism, are obstacles to life, to existential significance.

A real-life example is taken from the work of Fanon (1952/2008), who wrote, "The black man has no ontological resistance in the eyes of the white man" (p. 90). In racism, black people are not regarded as persons, are not singular beings in the eyes of racist individuals. Only persons, only singular beings, possess ontological resistance. Persons of color, such as Fanon, fall under the heading "black people," which denies their singularity, diminishing their dwelling in the world. Fanon recalled that a soldier, crippled in the war, "tells my brother: 'Get used to your color the way I got used to my stump. We are both casualties.'" Fanon's response: "Yet, with all my being, I refuse to accept this amputation. I feel my soul as vast as the world, truly a soul as deep as the deepest rivers" (p. 119). Fanon affirms his singularity and, in so doing, expresses his ontological resistance. He then becomes an obstacle to racists who, like the Borg, seek to force persons into the realm of the same.

I want to add to this discussion regarding obstacles and resistance. When Phillips (1993) indicates that the infant encounters the good-enough parent as an obstacle, he contends that this encounter of resistance gives rise to consciousness. Of course, this is not a reflective consciousness or knowledge evident later in psychosocial development. This early consciousness implies semiosis[5] and, as noted above, an attendant type of knowledge of the object and oneself. I will develop this in the section below, but for now we can consider that obstacles and resistance give rise to consciousness, knowledge, and a sense of self and other. This does not mean that knowledge is accurate, as is evident when Fanon (1952/2008) encounters the racism of white persons or what W. E. B. Du Bois (1903/2003) called the "double-consciousness" in response to the obstacles of racism. Generally speaking, the aims of this knowledge in relation to the object or obstacle are survival and flourishing, whether real or imagined. In terms of the encounter with the good-enough parent as obstacle, the infant is working to find knowledge of and ways to dwell resonantly with the obstacle—that which is other than the infant's sense of being-in-itself and being-for-itself. The parent's sense of being-in-itself and being-for-itself by definition resists the infant's singularity. In other words, the infant's knowledge and consciousness of self and other depend on resistance, which makes possible resonance in dwelling together. In a clinical setting, therapists are obstacles in the sense that they resist patients' singularities and vice versa. This is not a defensive resistance but rather an existential fact of singularity and the knowledge that arises in encounters with other singularities.

As noted above, the infant is also an obstacle, resisting by way of being-in-itself and being-for-itself. This resistance can be understood in terms of power. For Han (2019), "Power is the capacity to be with oneself in the other. … The power of the living manifests itself in the fact that it does not lose itself in the other" (p. 45). To lose oneself in the other would be to enter the world of the same, which obliterates singularity (annihilation or engulfment anxiety). For Han, power is part of agency, and agency resists being assimilated. More positively, infants' nascent agentic power makes it possible for them to resonantly be themselves in and with their parents. The Borg and zombies possess power, but it is a negative power that obliterates semiosis and, therefore, singularity, agency, and resonance. What is terrifying about the Borg, zombies, racists, and speciesists is that they represent annihilation of the power of being-in-itself. They exercise negative power—negation of singularity's existential resistance. Yet, many of us are fascinated by the Borg and zombies because, in my view, we are secretly assured of the impossibility of the Borg and, more to the point, assured of our own resistance that establishes our singularity and positive power.

All of this suggests that we can evaluate resistance and obstacles on a continuum of positive or negative. Positive resistance or obstacles are positive to the degree that persons' singularities are acknowledged and affirmed, which implies their resonant and relational survival and flourishing. From the other side of a mutual relation, positive resistance manifests the exercise of positive power and agency, giving rise to knowledge of self and other that both affirms and instantiates resonant dwelling together—being-for-and-with-others. By contrast, negative power denies or attempts to destroy singularities—leading toward death and sameness. This negative pole represents the eclipse of personal knowing and resonant dwelling together. Negative obstacles, as noted in the works of Du Bois and Fanon, can be internalized, giving rise to double consciousness and negative or dissonant embodied self-knowledge (see Coates, 2015; Laymon, 2019). In between these poles, there are varying degrees of being alive and dwelling. Utopias, for instance, represent ideal communities where there are a multiplicity of singularities who flourish as they dwell together. Naturally, none of us live in Edenic societies, but, in healthy or good-enough societies, mutual personal recognition and care affirm singularities and existential resistance while aiming for some degree of shared resonant flourishing. Toward the other end are dystopian communities, wherein individuals survive but do not thrive. Dystopian "communities" represent a resistance that manifests negative power and agency and dissonant dwelling. Of course, the Borg and zombies represent societies where community is not possible because everyone is the same—the living dead. Closer to home, a racist society, such as the United States, reveals the presence of a preponderance of negative obstacles, resistance, and power, giving rise to distorted consciousness (knowledge of self and other) and dissonant relations. It is an indecent society (Margalit, 1996).

Before addressing two other aspects of obstacles and resistance, it is important to return to the idea of resistance as representing a psychological defense, which is a dominant view in psychoanalytic literature. There are clearly examples in therapy and life where individuals resist acknowledging some aspect of themselves or certain painful memories. I think it was Carl Jung who said that it is a terrible shock to become acquainted with oneself. This suggests that resistance is a ubiquitous feature of individual and communal life. Yet, clinical resistance as a defense is usually connected to painful memories that patients do not want to re-experience or experience in the presence of the therapist. The greater the past trauma, the more defenses are deployed to avoid it. The patient, in one sense, is their own obstacle, and a good-enough therapist is an obstacle to a life magically lived without having to face and work through past wounds. The therapist is an obstacle to the desire for a savior.

But resistance is more than a defense in therapy. As mentioned above, the very existence of singularity implies existential resistance. So, both therapist and patient resist, but this is not necessarily defensive, simply an existential fact. There is also resistance as defense in the face of therapists who do not understand or who collude with systems that have marginalized and oppressed patients. A white male therapist's unconscious, racial biases may evoke defensive resistance in patients—women and people of color. Therapists may subtly or overtly suggest a patient of color or a woman learn to live with their "amputation," only to be faced with the patient resisting: "with all my being—I refuse to accept this amputation" (Fanon, 1952/2008, p. 119). In these cases, therapists are witting or unwitting obstacles that evoke the existential resistance linked to the patient's singularity. If this is labeled a defense of the patient, then it is a defense that aims to exercise positive power and agency that will overcome or render inoperative the obstacle of racism. But we can also label the therapist's interactions as an unconscious negative defense aimed at maintaining white male privilege.

Of course, resistance is not simply a clinical reality, whether we understand resistance as defense or not. As mentioned in Chapter 1, Aristotle argued that human beings are political animals. We are social and communal creatures, which means that, for the polis to survive and thrive, there must be mutual personal recognition and cooperation. Similarly, there must also be some level of civic care and trust. What does this mean in terms of obstacles and resistance? In a good-enough society, members must be good-enough obstacles for people to develop and possess senses of being-in-itself, being-for-itself, and being-for-and-with-others—singularities that are individual and relational. Put differently, civic care and trust provide relations wherein the singularities (and attendant resistances) of members are respected and the potentialities of societal members can be actualized. This is analogous to the care of the good-enough resistance of parents who facilitate infants' actualization of their potentialities—expression of their singularities. The presence of classism, racism, sexism, and so on, in a society indicates that cooperation and care are restricted to one group; both are denied to the group deemed to be inferior. Put another way, the dominant group exercises a kind of negative political, economic, and cultural power that is a negative obstacle to oppressed members of the society—undermining, in other words, their senses of singularity and flourishing.

Agency and the exercise of positive power points to another feature of resistance and obstacles, namely, freedom. Above, I mentioned Fanon's view of the goals of therapy, which are to raise to consciousness the real sources of suffering so that patients can choose an action toward these sources. There is an underlying liberative, political dimension to this view of therapy, which I understand as representing a process of facilitating patients' freedom (1) *from* the social-political apparatuses that chain their psyches and restrict assertion of their singularities and (2) *to* choose not only actions toward these apparatuses or negative obstacles but also actions that manifest their singularities—enacting existential resistance. It is not simply defending against these apparatuses but resistance as an expression and affirmation of individual and collective singularities. This extra-clinical or politic resistance aims at liberation, and liberation, in part, represents the affirmation of the singularities of marginalized and oppressed persons. Freedom, in other words, represents not simply self-efficacy but also the exercise of agency and power to actualize the potentialities of singular individuals and communities.

There is another feature of resistance and freedom of human beings. Freedom can be conceptualized in terms of an individual, but this is only part of the story. In Chapter 2, I introduced Hartmut Rosa's (2020) notion of "resonance," which he argued "is constitutive not only of human psychology and sociality, but also of our very corporeality, of ways we interact with the world tactilely, metabolically, emotionally, and cognitively" (p. 31). Implicit in his discussion is a relational notion of freedom that emerges from cooperation with other human beings and other species. Embodied, relational resonance is contingent on cooperation, which involves the shared recognition and acceptance of the existential singularity (and, hence, resistance) of the other. This cooperation and this recognition entail the exercise of positive power, which can be enacted only when there is relational or interpersonal freedom associated with resonant dwelling together. As Rosa notes, this "basic mode of vibrant human existence consists not in exerting control over things but in resonating with them" (p. 31). Relational dissonance, by contrast, emerges from the exercise of negative power as control over the other, denying not only the singularity of the other but also the cooperation necessary for the exercise of their freedom. The apparatuses of the ontological rift produce dissonance between human beings of the rift and othered human beings, other species, and the Earth, which is the focus of the next section.[6] This dissonance is evident in the manic desire to control nature, which I will return to below.

To summarize, the concept of resistance developed by Freud and others tends to be framed as a defense that interferes with the analytic process, which initially referred to the patient and later included the analyst. The sources of this interference can be intrapsychic and/or interpersonal. Patients' resistances are founded in fears and anxieties associated with past traumas being remembered, re-experienced, or re-enacted. While these empathic negative appraisals of resistance are helpful clinically, some later analysts have considered more positive evaluations of resistance—or at least some forms of resistance. I turned to Adam Phillips's discussion of obstacles in human life to tease out other extra-clinical features of resistance, which will be further developed in the section below on psychosocial development and eco-resistance. That said, I argued that an individual's singularity is itself a form of existential resistance to the singularities (obstacles) of other singular beings. This existential resistance is not necessarily defensive and broadly concerns the individual's resonant dwelling in the world of other singular beings. Also, resistance toward good-enough obstacles gives rise to consciousness and the exercise of positive power and agency, which are connected to freedom and the personal cooperation associated with resonant dwelling together. As noted, obstacles can be understood to exist on a continuum between positive life-affirming obstacles and life-denying obstacles. This continuum, in my view, can be used to appraise resistance in clinical or extra-clinical contexts.

The Rift, Resistance, and Dissonant Dwelling

In this section, I explore the idea of resistance that emerges in conjunction with Western apparatuses that produce the ontological rift. These apparatuses represent negative obstacles for othered human beings, other species, and the Earth. In addition, I consider how these apparatuses produce psychosocial defensive resistance, mainly against the reality of existential insignificance. Resistance as negative obstacles and defense points to a core feature of Western subjectivity, namely, the dissonance in dwelling in the world with other species.[7] This dissonance, I argue, is evident in Western apparatuses of philosophies and theologies that produce the rift and dissonance, which is a core feature of Western subjectivity. This existential dissonance is observed in the obsessive preoccupation with colonizing—subjugating, exploiting, and controlling—othered human beings, other species, and the Earth. More subtle examples of resistance as dissonant dwelling are found, in part, in the presence of antagonistic relations with the abstraction "Nature" and, closer to home, the absence of other species in psychoanalytic theory and practice.

A brief reminder of some of the features of the ontological rift is important. The apparatuses of the rift include the denial of personhood and, therefore, singularity to other species (and othered human beings). This results in a split between personal and instrumental, objectifying epistemologies. Other species, if they have any significance, have mere use value. That is, instrumental epistemologies affirm the putative superiority of human beings and inferiority of other species, which legitimates the use of other species by human beings captive to the rift. Instead of cooperation and reciprocity with other species, there is dominance, control, and exploitation. Recall also that other species are excluded from the polis (except for those instances where they are of use) and, therefore, they exist in the zones of non-justice and carelessness.

The ontological rift is a lived reality between human beings and other species. As noted in Chapter 1, those of us who have internalized and live out the rift "are structurally compelled (from without) and culturally driven (from within) to turn the world into a point of aggression. It appears to us as something to be known, exploited, attained, appropriated, mastered, and controlled" (Rosa, 2020, p. 14). We have seen this in Freud's writings, even though he rightly lamented the rift between human beings and other species. Our aggressive stance can be further understood in terms of resistance to negative obstacles. That is, by internalizing the apparatuses of the rift, we function as negative obstacles in relation to othered human beings and other species. Objectifying and exploiting othered human beings and species mean establishing relations where potentialities cannot be actualized. Functioning as negative obstacles also means that consciousness is distorted and freedom and agency are eclipsed or attenuated, which is more clearly observed when human beings are objectified and exploited. Yet, as documented in Chapter 3 regarding trauma, we can also take note of the damage negative obstacles of the rift have in relation to other species to whom we attribute sentience and with whom we identify, such as pets or Buck in Jack London's (1903/1981) story. Those who exploited Buck denied Buck's singularity, and as a result their knowledge of Buck and themselves was distorted. They resisted doing the work of knowing and caring for Buck. At the same time, Buck's consciousness of himself and of human beings was distorted by objectifying relations, whether these were relatively benign, at the homestead in California, or horrific, at the hands of his tormentors. When Buck experienced the love of John (a good-enough obstacle), he developed an ontological sense of self-significance and actualized his potentialities for freedom.

I add here that we, as negative obstacles, also overlook, deny, or overpower the existential and defensive resistance of other species. As noted earlier, all species possess singularity, which means, by definition, there is resistance to existential insignificance. Negative obstacles, in depersonalizing other species, overlook or deny this resistance, which is evident in the construction of other species as mute and dumb. When species are able to resist defensively, like Buck, subjects of the rift exercise negative power (violence and force) to overcome psychological defenses and existential resistance and, if continued, destroy the object—annihilating both defensive and existential resistance. Buck was viciously beaten because he resisted the negative power of his abusers.

So far, I have addressed negative obstacles of the rift in relation to other species. Those of us of the rift are obstacles to the significance and well-being of other species, evoking varied forms of species' resistances to our attempts to control and subjugate them. A question remains about the resistance manifested by negative objects themselves. What do we, as subjects of the rift, resist? Part of the answer was offered in the previous chapter regarding transference as projecting existential insignificance onto other species and the Earth. Resistance, like transference, is *a semiotic phenomenon of dwelling*. Western subjects' dwelling in the world represents a resistance to facing and accepting the reality of our existential insignificance (and impermanence) that is at the base of all semiotic valuations. I believe there is evidence for this claim. Freud (1917/1955c) argued that there have been three blows to human narcissism. The first shock was Copernicus, when human beings learned that we are not the center of the universe. The second narcissistic jolt was Darwin, when we learned that human beings are part of evolution, like all other living creatures. Freud (1917/1955c) added to the list, contending that human "megalomania will have suffered its third and most wounding blow from psychological research of the present time which seeks to prove to the ego that it is not even master in its own house" (p. 285). As far as I can tell, most Western human beings have sloughed off these blows with considerable ease, leaving much of life unchanged and raising the question of resistance. If megalomania is understood as beliefs in human superiority and dominion, then a defensive type of resistance emerges in relation to the rift and the attendant assumptive worlds it creates and maintains. That is, we ward off those narcissistic blows by hiding within the semiotic fortress we create to keep existential insignificance outside the walls where other seemingly mute and dumb species live and die. Those of us inside the walls of the fortress do not hear the cries from outside.

This resistance, however, is currently facing an obstacle it cannot ward off—the climate polycrisis as a mass extinction event. This does not mean that those of us in the semiotic fortress will have an epiphany and cede our resistance. Defensive resistances in the climate crisis, in other words, will become more frantic, hostile, violent, and destructive. The walls may be reinforced. Consider the rise of authoritarian leaders and parties around the world. Observe the obsessive, opportunistic lust for power for power's sake, revealing a nihilism that is basically a "Fuck you" aimed at othered human beings, other species, and the Earth. In the words of philosopher Wendy Brown (2023), "Nihilism is manifest today as ubiquitous moral chaos or disingenuousness but also as assertions of power and desire shorn of concern for accountability to truth, justice, consequences, or futurity, not only ethics" (p. 11). Political nihilism, in my view, is a symptom of resistance—a resistance to facing existential insignificance.

Consider also that, in the United States, climate agendas are ignored in favor of "Drill, baby, drill"—a celebration of the god of capitalism. Many Christian evangelicals and other religious conservatives who proclaim their love of Truth also manifest unconscious nihilistic angst in their support of capitalistic fascism and its attendant culture wars against marginalized peoples. All that matters is one's place in the kingdom of God, which means that the material world, other species, and othered human beings have no ontological significance except their utility in serving as a means for a select group of human beings to reach their fantasized aims. We are not simply facing climate deniers but persons who grasp blindly and tightly onto the beliefs in human superiority and dominion, resisting the climate crisis's revelation of our existential insignificance.

Resistance born of the rift can be further depicted in terms of dissonant dwelling in the world, which is, at times, evident in clinical settings. I want to first argue that existential dissonance has been depicted by Western philosophers and theologians for millennia, often under the heading of homelessness or restlessness. Plato's anthropology portrayed the dissonance in Western subjects, and, although this may be an overgeneralization, Western philosophers and theologians followed suit. The story of the cave reveals our resistance to becoming aware of the truth that what we have been seeing are mere illusions (Plato, 1956, *The Republic*, Book VII). Apparently, we prefer the illusions to the blinding light of ideal forms. There is, in other words, a dissonance at the very core of the human psyche. Plato's notion of ideal forms (*The Republic*, Books V–VII) was taken up by philosophers and theologians long after Plato had receded into existential insignificance. The varied inhabitants of the material world fall short of their ideal forms, suggesting a sense of not being fully at home in the world—a dissonance between the ideal form and its pale imitation. Human beings, in other words, can never measure up to their ideal forms. This strikes me as an anthropology rooted in shame versus a Christian anthropology based on guilt for having chosen to overlook the sovereign God's expectations and rules. In either case, for subjects of the rift, dissonance is at the core of the human psyche, and resonance becomes an epiphenomenon of ontological dissonance.[8]

A sense of being thrown into the world in search of home is evident in Jewish and Christian scriptures. In the origin myth, Adam and Eve are expelled from their Edenic home (ideal form) because of disobedience, leading to a life of existential separation from God and restlessness among their descendants—a restlessness or homesickness that is dispelled only when one resides in God or returns to God and God's kingdom (Augustine, 1963, p. 333). This tension between home and homelessness is reflected when Augustine, in the opening lines of his book *Confessions*, states, "You have made us for yourself, and our hearts are restless until they can find peace in you" (p. 17). For Augustine, it is not simply sin that creates alienation or homelessness; the very act of creation separates us from the Creator. For Augustine, like Plato, the Earth is a temporary abode or way station to the kingdom of God. This perennial theme of homelessness is evident in Western political theologies but is also echoed in political philosophies.

Jumping to the 20th century, two influential philosophers, Martin Heidegger and Emmanuel Levinas, agreed about existential homelessness but disagreed on its causes and implications. Heidegger, Gauthier (2011) argues, "interprets homelessness as a symptom of the abandonment [or forgetfulness] of Being by beings" (p. 129), pointing to its universal ontological reality. Heidegger, then, viewed his philosophy as a response to homelessness. His position is reminiscent of Georg Philipp Friedrich Freiherr von Hardenberg's (pen name Novalis) belief that "philosophy is really homesickness—the desire to be everywhere at home."[9] Of course, Heidegger eschewed the idea of everywhere and opted for place, identity, and tradition to mitigate our existential estrangement. Heidegger believed, Gauthier (2011) argues, that the challenge "is to create a philosophy that will facilitate a return to rootedness to help man to become at home in the world" (p. 9). "I know that everything essential and everything great," Heidegger writes, "originated from the fact that man had a home and was rooted in tradition" (as quoted in Gauthier, 2011, p. 92). Critics of Heidegger argue that his obdurate focus on place, identity, and tradition in relation to existential homelessness fits well with the Nazis' racism, antisemitism, and German exceptionalism—iterations of the rift. Phillip Tonner (2018) also criticizes the Heideggerian philosophical view that "only anatomically fully modern humans dwell" (p. 9), which disavows the dwelling of early humans as well as of all other species.

Emmanuel Levinas's philosophy is, in one sense, a critical response to Heidegger. Levinas contends that Heidegger "effectively subordinates ethics to ontology" (Gauthier, 2011, p. 104) and that it is ontology that tends to eliminate "the alterity, or otherness, of the comprehended being" (p. 105). More pointedly, Gauthier states that, in Levinas's view, "The Heideggerian project is ethically problematic because it is oblivious to the needs of strangers" (p. 97). Nevertheless, Levinas accepted the premise of human homelessness, though he believed that the presence of others and our refusal to be obliged to the Other are why human beings experience homelessness (p. 113)—rather than the forgetfulness of Being. For Levinas, the source and sense of homelessness "provide[s] the impetus for human fraternity" (p. 114). Gauthier notes that, for Levinas, "the home achieves its full dignity when the Other is welcomed" (p. 131). In his book *Totality and Infinity*, Levinas (1969) states,

> To dwell is not the simple fact of the anonymous reality of a being cast into existence, as a stone one casts behind oneself; it is a recollection, a coming to oneself, a retreat home with oneself as in a land of refuge, which answers to a hospitality, an expectancy, a human welcome.
>
> (p. 156)

Here is the apparent paradox of Levinas's view: the Other evokes a sense of homelessness, and this homelessness is mitigated by the infinite obligation to recognize and welcome Others in their singularity.

Levinas is particularly helpful here because of his relevance for the discussion below. For now, I simply wish to acknowledge that, in philosophical and theological literature, the senses of dwelling, belonging, homelessness, and homesickness have been themes for millennia. I understand the theme of homelessness as a symptom of the ontological rift, one that manifests a dissonance at the core of Western subjectivities. Put differently, apparatuses of the rift create a dissonant relation between human beings and what we abstractly construct as Nature, which becomes an object or obstacle we must resist. Western subjects are not at home in the world. We do not live or dwell resonantly in relation to other species and the Earth.[10] Put differently, embodied, relational resonance is something that previously existed in Eden or will exist in the kingdom of God or in ideal forms. Naturally, we may obtain glimpses of this in life, which means resonance is an epiphenomenon. All of this is philosophically and theologically essentialized, as if it is a universal and perennial feature of human existence. All the ink spilled in talking about human existential homelessness is itself a form of resistance—resistance to acknowledging and accepting the reality of our existential insignificance vis-à-vis all creation.

Is there evidence of this essentialized dissonant resistance in Western history? Historians reveal and sometimes justify innumerable stories of people being unhoused—killed, colonized, enslaved, and so on (e.g., Danner, 2009; Harman, 2017; Lepore, 2018; Zinn, 2003). Orlando Patterson (1982), Charles Mills (2017), and Edward Baptist (2014) describe the horrific realities of U.S. slavocracy and the Jim Crow era for African Americans who were denied social-political dwelling, whether literally by being killed or by being deprived of the resources necessary to thrive. Feminists (e.g., Fraser, 2020; Tronto, 2013) and other liberation philosophers and theologians (e.g., Cone, 1970; Dussel, 1985; Gutiérrez, 1985; Moltmann, 1973) have long pointed out how excluded-included Others are marginalized and oppressed, which means being pushed to live on the fringes of society where they are deprived of the goods necessary for their well-being—for dwelling. Sociologists, such as Mathew Desmond (2016, 2023), depict the deeply painful and disturbing realities of people who are legally (but unjustly) and forcibly evicted from their homes,[11] which often leads to homelessness in a country that has the resources to create homes for everyone. Joe Soss et al. (2011) and Loïc Wacquant (2009) provide comprehensive sociological analyses of the precarities of poverty, exposing the systemic realities that leave millions of people housing- and food-insecure. Housing and food insecurity are linked to systemic apparatuses of racism and classism, which undermine the dwelling of persons of color and those who are constructed as lower class. Georg Lukács (1968), Herbert Marcuse (1964), Dany-Robert Dufour (2008), Wendy Brown (2015), Nancy Fraser (2022), Nancy Fraser and Rahel Jaeggi (2018), Jerry Mander (2012), Sayak Valencia (2018), and others address the political and economic realities resulting from global capitalism that lead to psychological, political, and economic experiences of alienation and homelessness—the unhousing of human beings and other species. As mentioned in previous chapters, factory farms, industrial slaughter factories on land and sea, and the destruction of habitats from mining have resulted and still result in the deaths of untold billions of other species. All of these examples illustrate our lack of living resonantly with other human beings, other species, and the Earth. This dissonance, I contend, is a symptom of the ontological rift and our obdurate resistance to facing the reality of our existential insignificance.

Some may note that killing and violence are evident in Indigenous ways of living, which suggests that dissonance is universal. I agree to the extent that human beings and other living beings can experience dissonance, whether that is in killing or being killed or in some other harm that interferes with one's dwelling in the world. Indeed, Indigenous narratives often portray relational dissonance between human beings, between human beings and other species, and between other species. There are also depictions of homelessness or restlessness. However, in my reading of hundreds of these stories, dissonance is a possibility of being alive, whether for humans or other species. Moreover, as a possibility, dissonance can also be remedied in the here and now. This means that dissonance is a feature of existence, but it is not an existential essence or central core of human psyches or human dwelling in the world. Indeed, many Indigenous stories illustrate the idea that resonant dwelling is the central feature of life and materiality, while dissonance represents something that goes awry, eliciting responses of repair.[12] Indigenous resonant dwelling in the world is the "basic mode of vibrant human existence" (Rosa, 2020, p. 31), whereas dissonance is an epiphenomenon.

Since I have mentioned Indigenous stories, this is also the place to elaborate further on the difference between Western defensive resistance to existential insignificance and Indigenous philosophies of dwelling in the world that affirm the existential significance of other species and the Earth—which includes some measure of accepting the uncontrollability of other species and the Earth. Western apparatuses of the rift manifest a preoccupation with controlling nature, revealing a defensive resistance to the reality of uncontrollability, which Rosa (2020) argues is a key feature of resonant dwelling with others. This unconscious fear of uncontrollability is linked to existential insignificance. That is, in exercising control over nature and other species, Western subjects obtain senses of self-efficacy and freedom tied to an existential significance that is dependent on the apparatuses of the rift. Uncontrollability, then, becomes linked to profound helplessness, as well as a loss of self-efficacy, freedom, and existential significance. Understandably, the very possibility of uncontrollability heightens anxiety, which is linked to our unconscious terror of our existential insignificance. Instead of getting in touch with both, we build the semiotic walls higher and deeper.

If Western defensive resistance is understood as avoiding the reality of existential insignificance by projecting this onto other species and the Earth, does this mean that Indigenous peoples accept existential insignificance? Or, better, how do they handle existential insignificance? One response is first to acknowledge that singularity and semiosis occur in relation to existential insignificance. Indigenous philosophies affirm the singularities of other species as well as of the land or Earth. In a sense, this means there is a recognition of the shared precarity and uncontrollability of all life and, if you will, the categorical command to care for the land and its inhabitants. This does not mean they consciously "accept" existential insignificance, but they do not defend against it. In other words, they do not project existential insignificance onto other species or secure their own significance in an exclusive heaven at the expense of the significance of other species and the Earth. Indeed, their cosmologies are populated by other species that have individual intelligences. Moreover, there is a consistent theme that their home is where they live. They certainly have experienced homelessness (ethnic cleansing) and dissonance, but home is a place in the here and now—a home shared with other species on this one Earth.

Let me offer an example to illustrate these claims. I return to Jonathan Lear's (2006) exploration of the Crow people's response to the cultural devastation wrought by settler colonialism. It was not simply that the Crow community had been forcibly removed from their land, which was bad enough since land was integral to their cultural traditions (Lloyd, 2024). It was also that these very cultural traditions were under assault. Plenty Coups, the Crow chief, had a dream or vision wherein a small bird showed his people how to survive the onslaught of Western colonizers by keeping their traditions. Naturally, Crow persons experienced profound sadness, grief, depression, and so on, as a result of their being dislocated and having their traditions undermined by white colonizers' attempts to force assimilation. To be sure there was great dissonance and profound experiences of homelessness. Yet, they did not respond with nihilism to what was happening to them. In my view, this was because their sense of existential significance was shared with other species and the Earth. Put differently, they may have lost their homeland, and their traditions were under assault, but their core beliefs that all living beings and the Earth have significance enabled them to survive. Existential significance was grounded in the Earth and its inhabitants. This central belief (and others) was also the source of their positive resistance in the sense of asserting their singularities in the face of oppression and marginalization.

By contrast, many religious persons of the rift have generally located the sole source of their significance in a sovereign God and only secondarily, at best, in the land or the Earth, and that kind of significance was and is usually instrumental. So, the rise of modernity and the death of God were and are psychologically and socially disruptive because their sense of significance no longer has a foundation, leaving Western persons open to depression, nihilism, and violence. Let me stress that the Crow people's significance was founded on the Earth and its inhabitants, not on a singular superior being. This means that their resistance to existential insignificance was not a psychosocial defense, though the affirmation of their singularity and the singularities of other species and the Earth represented positive resistance in the form of affirming their own existential significance, as well as that of other species and the Earth. There were, of course, experiences of dissonance in their lives, but dissonant dwelling was not a central feature of their anthropology or cosmology. Rather, the central feature was and is resonant dwelling with other species and the Earth.

Reimagining Resistance: Psychosocial Development, Inoperativity, and Eco-Resistance

In this section, I argue that a psychoanalytic view of resistance can be emended to include forms of resistance that manifest resonant ways of dwelling together, which I suggest is evident in Harold Searles's (1960) and Susan Kassouf's (2017) invitations to include other species in our understanding of psychosocial development. In addition, I contend that, like eco-transference, humans can experience eco-resistance, which is existential in the sense of affirming singularities and multinaturalism and, at times, defensive. Eco-resistance as a defense can take the form of political resistance in rendering inoperative the apparatuses of the ontological rift—a positive form of resistance.[13] This inoperativity accompanies collective political affirmations—cognitive and actional—of the singularities of other species and the Earth. Furthermore, eco-resistance creates spaces to include other species in discourses regarding care and justice. To depict eco-resistance and its features, I return to human psychosocial development.

As noted in previous chapters, long before symbolization and propositional knowledge, infants engage in early semiotic organizations of experience, even prior to birth. This living being prior to birth, like all living beings, affirms its singularity (being-in-itself, being-for-itself) semiotically and, therefore, resists the absence of semiosis—sameness and death (existential insignificance). This is the earliest and most basic form of existential resistance—a resistance born not as a psychological defense against existential insignificance but out of affirmation of singularity—being-in-itself, being-for-itself. But this early nonpropositional semiotic affirmation occurs in the context of a relation(s) and embodied experiences of resonant dwelling in the womb and, later, with parents.

Once born, infants' good-enough parents' personalizing and caring attunements, in general, represent the affirmation of the infants' singularities. More particularly, these attunements to infants' assertions are necessary for actualizing infants' potentialities, including more complex semiotic organizations. This also means that parents' care entails an acceptance and affirmation of infants' existential resistance, as manifested in their singularities. That is, the intersection of care and the actualization of potentialities accompanies semiotic constructions of shared singularities—being-in-itself, being-for-itself, and being-for-and-with-others—which are conceptualized as embodied, relational resonances.

Of course, early forms of embodied, relational resonance may contain experiences of dissonance. It is easy to imagine infants' experience of dissonance at birth, as well as during moments of hunger or distress. Good-enough parental caring responses to infants' cries, which include repairing parental failures, are attempts to restore resonant relations. When successful, infants learn that moments of dissonance between parents and themselves can be managed and resonance restored, which deepens nonpropositional relational trust and hope.

Infants' presymbolic organizations of trust and hope can also be understood in terms of uncontrollability associated with parents' and infants' resonant relations. There is a nascent agency in early life, and this agency is in dialectical relation with the reality that one cannot control parents (obstacles) or the environment. Parents' good-enough attunements provide infants with a sense of trust that uncontrollability is not simply tolerable but a feature of resonant relationships. It is this uncontrollability and this trust that make possible an openness to something new, to creativity, to the possibility of resonant relations with other objects.

All of this, then, is not simply in relation to the parents. In the previous chapters' discussions of primary transitional objects (PTOs), I indicated that PTOs represent organizations of experiences of child–parent interactions. To infants, the world of objects is alive, and the particular PTO represents the aliveness and resonance (including the repair of dissonance) experienced in relation to their parents. Moreover, the PTO is a singular alive object that by definition resists, and this resistance instantiates uncontrollability in the context of a trusting relationship. Also, in a child's animistic imagination, the PTO is a singular alive object that affirms the singularity of the child, establishing a relational resonance of beings who are in and for themselves with another—solidarity. This is a shared positive resistance to the realities of existential insignificance—sameness and death.

Understandably, parental deprivation undermines infants' actualization of potentialities, evoking a more defensive hold on semiotic constructions of being-in-itself and being-for-itself—singularities. In addition, parental deprivations and impingements heighten children's distrust and fear of uncontrollability, which is joined to embodied, relational dissonance. Stated differently, deprivation evokes resistance toward uncontrollability manifested in psychological defenses to maintain and control whatever resonant experiences the child has. Resistances in the wake of parental deprivations can also be understood as representing not only the desire to survive but also the hope, real or imagined, of achieving some sense of embodied, relational resonance. The PTO, on these occasions, may be used defensively to provide children with a sense of resonance amid parent–infant dissonance.

As children begin to actualize their potentialities for varied types of symbolization, the secondary transitional object (STO) makes its appearance—an object that is chosen from and represents the cultural field. As depicted in previous chapters, the STO initially represents the intersection of animistic and personal epistemologies. The child is able to recognize the STO as not only alive but also as a person. To return to the comic strip *Calvin and Hobbes* (Watterson & Trudeau, 1985–1995), Calvin imaginatively recognizes Hobbes as a tiger-person, and Hobbes recognizes Calvin as a human-person. While there is an embodied, relational resonance between them, there are also moments of dissonance that are repaired. Existential resistance is represented in their mutual affirmation of their distinct personhoods and shared resonance. There is no individual or shared defense or negative resistance here—only resonant solidarity.

Ideally, this shared resonance and this existential resistance are eventually extended across the cultural field, which, in Indigenous contexts, includes human and other-than-human persons as well as the land. Indigenous communities do not separate culture and nature or politics/civilization and nature. These are Western binary abstractions that represent dissonant relations and, as indicated above, negative forms of psychosocial defensive resistance against coming to terms with humans' existential insignificance. This does not mean that Western subjects cannot experience resonance among themselves, but they have greater difficulty in experiencing resonance with other species and the Earth because they have internalized the apparatuses of the ontological rift and its attendant psychological defenses (rationalization, denial) against existential insignificance. Consider, for instance, the

> overwhelming evidence that children are healthier, more resilient, and more creative, and have higher expectations of self-efficacy, when they are in regular contact with animals and expansive natural spaces in which they experience sensual/practical resonant relationships, the therapeutic value of which for adults has also been well documented.
>
> (Rosa, 2019, p. 272)

Rosa's claim implicitly suggests that there are children and adults who are not healthy because, in my view, they are operating out of the apparatuses of the rift. Although it is also possible for Western children and adults (including scientists) to have nondefensive, resonant relations with other species and the land, the apparatuses of the rift must be rendered inoperative for these experiences to emerge.

The above leads me to address what I call eco-resistance, which can be understood as an affirmation of the singularities of living beings, an acceptance of multinaturalism, and, at times, a psychosocial positive defense—a form of political resistance rendering inoperative the apparatuses of the ontological rift. This eco-resistance creates spaces to include other species in discourses regarding care and justice. Eco-resistance, for instance, is found in Indigenous epistemologies in the sense that they affirm the singularities of other species and the Earth. There is, in other words, a shared recognition that resonant dwelling together is a feature of being alive in this place upon which all beings depend. This is a shared existential resistance in relation to the realities of existential insignificance and impermanence that lie at the basis of semiosis and life. I add that eco-resistance creates spaces for embodied, relational resonances with other human beings, other species, and the Earth, as well as spaces for managing and repairing, if possible, moments of dissonance.

Eco-resistance by Western subjects of the rift can also operate as a positive political defense. By this I mean that persons who actualize their capacity for impotentiality or their capacity to render the apparatuses of the rift inoperative are defending against living out of these apparatuses even as they have internalized them. More specifically, in rendering the apparatuses of the rift inoperative, they aim to secure shared or collective experiences of resonance between themselves and, for Indigenous peoples, other species and the land. Using a different concept, their capacities to differentiate from these apparatuses and to choose to affirm the singularities of living beings are positive, nondefensive eco-political resistances aimed toward establishing embodied, resonant relations with other species and the Earth. In so doing, these political eco-resistances affirm the uncontrollability of life as well as the categorical command to care (anarchically), creating spaces of inclusion and justice. A few examples will illuminate these claims.

In Chapter 1, I noted that Ruby Sales, an African American activist, grew up in the Jim Crow South with its racial contract.[14] Recall that Ruby said,

> [My] parents were spiritual geniuses who created a world and a language where the notion that I was inadequate or inferior or less-than never touched my consciousness. I grew up believing that I was a first-class human being and a first-class person, and our parents were spiritual geniuses who were able to shape a counterculture of black folk religion that raised us from disposability to being essential players in society.[15]

Ruby's parents' anarchic care for their children can be understood as political resistance, rendering inoperative the apparatuses of racism's ontological rift and creating spaces for shared experiences of resonance between themselves and the children within their community. The parents' care can also be understood, in part, as a positive, defensive eco-resistance in the sense of shared dwelling resonantly together.

Indigenous peoples also suffered because of the Euro-American apparatuses that produced and produce the ontological rift (settler colonialism) between themselves and Indigenous peoples. Chief Seattle, in a speech addressed to the U.S. president in 1854, said:

> Our dead never forget this beautiful world that gave them being. They still love its verdant valleys, its murmuring rivers, its magnificent mountains, sequestered vales and verdant lined lakes and bays, and ever yearn in tender fond affection over the lonely hearted living, and often return from the happy hunting ground to visit, guide, console, and comfort them. … Every part of this soil is sacred in the estimation of my people. Every hillside, every valley, every plain and grove, has been hallowed by some sad or happy event in days long vanished. Even the rocks, which seem to be dumb and dead as they swelter in the sun along the silent shore, thrill with memories of stirring events connected with the lives of my people, and the very dust upon which you now stand responds more lovingly to their footsteps than yours, because it is rich with the blood of our ancestors, and our bare feet are conscious of the sympathetic touch.[16]

Chief Seattle was, in one sense, acknowledging the resonance his people had with each other, other living beings, and the land, while also pointing to the dissonance wrought by Christian settler colonists. This dissonance was not and is not simply between Indigenous persons and colonizers but is also between colonizers, other species, and the Earth.

Similarly, John Neihardt (1932/2014) wrote down the story of Black Elk and his visions. In this story, Black Elk details the struggles of his people, the Oglala Lakota, in the face of the violent incursions of white people onto their lands. Black Elk, after finishing his story, exclaimed,

> Grandfather, Great Spirit, once more behold me on earth and lean to hear my feeble voice. You lived first, and you are older than all need, older than all prayer. All things belong to you—two-leggeds, the four-leggeds, the wing of the air and all green things that live. … Day in and day out, forever, you are the life of things.
>
> (p. 171)

On the one hand, Black Elk's prayer represents an affirmation of the embodied, relational resonance he and his people had with each other, other species, and the Earth. In other words, "Great Spirit" is an abstraction of this resonant dwelling in the world that has not been annihilated by white persons. This affirmation, like that of Chief Seattle's, is a form of political eco-resistance—resisting and rendering inoperative the apparatuses of the ontological rift. It is a positive, defensive resistance.

Lest we think settler colonialism ended in the 19th century, Dana Lloyd (2024) details the recent struggles of Indigenous peoples to secure their way of life and the land in the face of the U.S. legal system. Lloyd mentions Turtle Mountain Ojibwe scholar Heidi Kiiwetinepinesiik, who writes, "Stories are law. … Recently, critical approaches to the study of Indigenous law have elucidated this point, noting stories lay out the central principles for how people order the world" (p. 28). This is echoed by John Borrows, an Anishinaabe legal scholar. He writes, "Law is us. And it's the animals, and it's our dreams, and it's our stories, and it's our relationships" (p. 31). The high country that various Indigenous peoples were seeking to protect included "the mountains, rivers, wildlife, and ocean … in which each part is related to each other part" (p. 42). Lloyd writes,

> Because the trees themselves are considered living beings, their removal would be irreligious. But as the trees are living beings, as part of the community, then their removal is an attack on the community, and it should be thought of as a continuation of the genocidal policy.
>
> (p. 43)

This is settler colonialism's ontological rift. Lloyd rightly stresses that "instead of anthropomorphizing nature, Native epistemologies open the category of personhood to include nonhuman beings" (p. 134). The various Indigenous groups manifest resonant dwelling that conflicts with the U.S. legal system, which is an apparatus of the rift. Their way of life and assertions of their ways of life are acts of political resistance—acts of eco-resistance.

In summary, resistance, in terms of early psychosocial development, can be conceptualized as an agentic affirmation of the singularities (nondefensive existential resistance) of other living objects, giving rise to resonant dwelling together between parents and children and between children and other objects. The emergence of symbolization and STOs further solidifies this solidarity, wherein there is an imaginative sense of the mutual affirmation of singularities. This is solidarity as shared existential resistance. Eco-resistance, I noted, represents the affirmation of the singularities and the attendant existential resistances of other human beings, other species, and the Earth. This fosters types of ecological knowledge and a consciousness of other species and the land, which can be understood as political resistance, wherein other species and the Earth are included in the polis, the categorical command to care, and the zone of non-justice.

Resistance and Therapy

All of this may seem far afield from the consulting room and the resistances manifested by both patients and therapists. I certainly want to retain the idea of resistance as anything that interferes with the therapeutic process or insight-oriented work. Yet, given the climate crisis, I advocate broadening our use of and understanding of the concept of resistance. I add that there is a liberative strain in psychoanalytic history, whether that is evident in patients such as Bertha Pappenheim or in the works of Frantz Fanon (Gibson & Beneduce, 2017; Laubscher et al., 2022), Gherovici and Christian (2019), and Layton, Hollander, and Gutwill (2006). This liberative strain points to the presence of positive resistance in therapy, whether that is cultivated in the process of therapy or manifested as a defense against the failures of therapists to understand and appreciate the experiences and motivations of their patients. Given this, I begin with how psychoanalytic theories and training can operate as sources of resistance against an awareness of the destructive ways we dwell in the world. I then consider how a reimagined positive resistance can shape theory and practice with some patients.

In Chapter 1, I argued that Freud, understandably, made use of Western systems of thought to construct psychoanalytic theories and, therefore, that these theories functioned, in part, as an apparatus of the rift between human beings, other species, and the Earth. This resulted in, I asserted, the formation of a normative, historical, and political unconscious when it comes to other species. Some human beings (not othered human beings) are deemed to be existentially normative (believed to be superior), meriting historical memories and political inclusion. As I have argued in this chapter, the apparatuses of the rift foster dissonant relations between other species and the Earth. This is also evident in Freud's view of nature, which is an abstraction revealing antagonistic, instrumental, and objectifying relations with other species and the Earth. These dissonant relations represent a type of defensive resistance against becoming aware not just of our dissonance but also of our fears and anxieties regarding our existential insignificance and impermanence. A critical reader may object, pointing to psychoanalysts such as Lacan, who seemed willing to embrace the "lack" at the core of human subjectivity, or to Winnicott, who, in the last two years of his life sought, to reduce his ego so as to enter the canal of death (LaMothe, 2014). I would counter with the evidence that psychoanalytic developmental theories, as Searles (1960) and Kassouf (2017) have pointed out, have excluded other species and the land and have tended to use the term "primitive" when addressing animism. This exclusion is a symptom of negative or non-ecological resistance. In addition, there is a dearth of literature regarding analytic practice that explores patients' beliefs, perceptions, dispositions, attitudes, and behaviors toward other species. Further evidence is noted in various websites of analytic institutes. Of the 34 training institutes in the United States, a few of them address, in one form or another, political concerns; none of them address ecological concerns, whether in their mission statement or syllabi (not all sites include syllabi). This absence is a symptom of resistance in the realm of theory, training, and practice.

To be fair, I believe a major part of this collective resistance may not simply be a defense against existential insignificance but rather reflects a resistance to change. Most therapists, because of their education, training, and personal biases, identify with and internalize a particular theory or school. In general, psychoanalytic training entails analysands being introduced to various schools of thought and attendant practices. In the process of internalizing and identifying with a particular school, analysands obtain a professional identity as well as a sense of meaning and purpose. A cursory reading of psychoanalytic history reveals the battles, sometimes vicious and usually unanalyzed, between psychoanalysts of different persuasions (Breger, 2000; Gay, 1988). So, resistance to changing our theories and training may simply reside in our routine cherished identifications.

Another ordinary and understandable source of resistance concerns the work of empathy and compassion. Fanon (Gordon, 2015) had a great deal of empathy and compassion for those who suffered oppression and marginalization at the hands of French colonizers. Yet, French therapists largely failed to heed Fanon's voice or the voice of othered human beings, in this case Algerians. In the United States, at around the same time (the 1950s), the Lafargue Clinic in Harlem cared for African Americans suffering the wounds of racism. Its clinicians, I suspect, had considerable empathy and compassion for the victims of a racist (and classist) society. However, racism as a topic for psychoanalysts was a rarity (searches for "racism" in the title in the Psychoanalytic Electronic Publishing library find 2 articles in the 1950s, 5 in the 1960s, 47 in the 1970s). This gradually changed in the 1990s (319; remarkably, since 2020, 1,170 articles with "racism" in the title are identified). This change represents, in part, the collective work of increasing empathy and compassion toward those who have been marginalized. In other words, to cross the rift and demonstrate empathy and compassion in relation to human beings who have been othered takes work. History is replete with examples of our resistance to do this work.

I suggest that a parallel resistance exists regarding the suffering of other species. First of all, consider that, in the 1990s, only three articles in the Psychoanalytic Electronic Publishing library mention climate change, and nine in the 2000s. Since 2010, over 600 articles and books mention this topic, representing a significant surge of interest in this area. Interest in the lives of other species has experienced a smaller surge. Since 2000, approximately 47 articles have addressed other species and climate change, with a number of authors concerned about the well-being of other species (e.g., Cambray, 2017; Clough, 2024; Funk, 2024; Gentile, 2018, 2023; LaMothe, 2023; Rendón, 2020; Turnbull & Bär, 2020). This smaller surge may represent a turning point toward developing greater interest, empathy, and compassion for other species, but, at the moment, the dearth of attention suggests defensive, negative resistance.

I recognize the psychosocial labor required to recognize and care for the singularities of other species. It not only includes the labor of revising our theories and adjusting our therapeutic practices; it also requires altering our dispositions and behaviors toward other species, especially those species that are not deemed to be cute or cuddly (e.g., insects). As mentioned in an earlier chapter, the philosopher who said she refused to read a book recommended by her friends because she wanted to continue eating meat represents this kind of psychological resistance to the work of compassion and empathy—resistance to crossing the rift.

Fair enough, a critic might say, but, at the risk of appearing resistant, how does all of this work when we consider that psychoanalysis and psychoanalytic training are aimed at trying to understand the suffering and struggles of human beings, which is challenging enough? What, in other words, does this have to do with training and practice? I have several responses to this question, even as I recognize that more collective conversations regarding training and practice are required. First, in one sense, each of us needs to ask ourself if our dispositions and behaviors toward other species are symptoms of resistance, and, if so, what are the sources of our resistance and what are we resisting? Quickly dismissing this question is akin to the resistance demonstrated by the philosopher mentioned earlier. Second, the belief that psychoanalysis is for human beings is correct, at least on its face. However, it screens two important existential truths, namely, that semiosis is a feature of all life and that human psyches are inextricably a part of the diverse habitat we call Earth. Only those captive to the rift believe human "minds" are somehow separate from the abstraction of "Nature" or that other species do not have minds or psyches.[17] The human mind is ecological in that it is situated in and dependent on a biodiverse Earth. This does not mean that we should psychoanalyze other species, especially given that we often do not do this very well with human beings. We could, though, adopt a listening stance in relation to their communications. More specifically, this means (1) addressing questions regarding the relationship between human psyches and the psyches of other species; (2) understanding patients' psychosocial development and that of other species; and (3) exploring clinically, when appropriate, patients' beliefs, attitudes, and behaviors toward other species and the Earth. This latter point does not mean the "use" of other-than-human species as part of a patient's analysis. Rather, it is about addressing questions of how we and our patients dwell resonantly or dissonantly in a biodiverse world of singular beings. In short, dealing with and working through resistance, our own or our patients', takes work.

So far, I have been addressing resistance as a defense, and a negative one at that. Above, I briefly provided illustrations of resistance as a positive defense. Let me continue with this theme. Some psychoanalysts (Altman, 2000, 2004; Aralepo, 2003; Benjamin, 1995, 1996; Cushman, 1995; Dalal, 2002; Grand, 1997; Kovel, 1970, 1984, 1988; Laing, 1969) have manifested positive political resistance in relation to the systems of oppression, namely, racism, classism, and sexism. Likewise, other analysts (Hoggett, 2012, 2013, 2019; Orange,

2017; Weintrobe, 2010, 2013, 2021) have persisted in expressing, in my view, positive political resistance to the apparatuses (e.g., capitalism) of climate change. We need not restrict the liberative dimension simply to human beings who suffer marginalization and oppression. Decades ago, philosopher Peter Singer (1975) argued for the liberation of other species. To aim at this liberation entails rendering inoperative the apparatuses of the ontological rift. This is not simply preferring not to operate out of the grammar of the rift; it also involves actions that affirm the singularities of other species and their inherent uncontrollability. To recognize the singularity of another being is to accept some degree of its uncontrollability, which is necessary for the possibility of experiences of resonance (Rosa, 2020).

The only way, then, that we can render these apparatuses inoperative is to become conscious of their pervasive operations in our lives. Becoming conscious of what was previously unconscious means we are placed in a position to decide to address the obstacles of the ontological rift. The philosopher who decided not to read the book was conscious of what the book contained but chose to live out of the apparatuses of the rift, which means that the liberation of other species did not matter to her. But another choice is possible, as is evident in Singer's works as well as in the works of some psychoanalytically educated authors and clinicians. This decision leads to varied individual and collective actions that render inoperative the apparatuses of the rift—actions I view as forms of eco-resistance. Naturally, as Giorgio Agamben (2004) notes, rendering apparatuses inoperative does not mean these apparatuses no longer operate. They continue to have their effects. However, those who together render these apparatuses inoperative avoid participating in perpetrating these negative effects.

However, I have not provided a clear answer(s) with regard to training and therapy. I have to acknowledge my limitations regarding my own analytic training and practice. Exploring my dispositions, thoughts, and feelings regarding other species was not a feature of my training or analyses. And I can think of only a few patients who have explored with me their relationships with pets, though this did not extend to other species or their thoughts and dispositions about the Earth. In the last six years, only three or four patients have expressed varied eco-emotions and thoughts about climate change, and only one mentioned their grief about species near extinction. Engaging in this exploration with some patients is a new territory that I believe will be increasingly important when people become more aware of how the climate crisis is impacting not only human beings but also millions of other species. With regard to the patient who mentioned their grief at the realization that so many species are close to extinction, I did not explore their views and relations to other species. This may be a manifestation of my own resistance, but it is also, I believe, associated with my being unaware of how to approach or handle this. Analytic therapists are trained to explore patients' past and present familial relationships, their sexual lives, their political views, and so on. It is not only that we, as Searles (1960) noted, need to include other species in our developmental theories; we also need to find ways to explore our own and (some of) our patients' experiences, beliefs, dispositions, and so on, regarding other species and the Earth. This and other chapters are aimed at contributing to the dialogue.

Conclusion

The idea of resistance in psychoanalysis is rooted in its very origins. It is a useful clinical concept for understanding the communicative dynamics between patients and therapists that interfere with the process of therapy. Clearly, resistance as a defense is also applicable to non-clinical situations wherein people employ psychological defenses to avoid becoming aware of what heightens their fears and anxieties. The climate crisis and the myriad ways people avoid facing it are examples of resistance as a negative defense, hindering (1) recognition of the sufferings of othered human beings and other species and (2) effective action aimed at helping other species, as well as individual and collective action toward mitigating climate catastrophes. This negative resistance, I have argued, stems from the internalization of the apparatuses of the rift, which gives rise to dissonant subjects who defensively ward off the fears and anxieties of their existential insignificance, evacuating them onto other species.

Yet, not all resistance is negative or fundamentally defensive. To be alive, to be a singular being, is to resist sameness and existential insignificance. This positive resistance can also be political in the sense of persons resisting political, economic, and social forces and social imaginaries that undermine the well-being of persons, whether "persons" refers to human beings or other species. Positive political resistance is also evident in rendering inoperative the apparatuses of the rift, which is accompanied by the possibility of resonant relations and experiences. This eco-resistance affirms the singularities and uncontrollability of other living beings, creating spaces of resonant dwelling together on this one Earth. These various types of resistance, I suggest, have clinical and extra-clinical relevance, which raises questions concerning analytic training and practice. My hope is that this and the previous chapters will spark more conversations about psychoanalytic theories and practices in relation to the climate crisis.

Notes

1 It is ironic that the "talking cure" developed as a result of a patient, Anna O (Bertha Pappenheim), who basically told Freud to shut up and listen (Breger, 2000, pp. 103–110). Freud obviously did not consciously see himself as resisting her, but to his credit he listened. One could say that Pappenheim's resistance encountered and overcame the doctor's resistance—a resistance that emerged from patriarchal social, cultural, and political apparatuses. Psychoanalysis, in part, was born when

the good doctor set aside his resistance for the sake of creating a space for the patient to appear as she was and not as what patriarchal society expected of her. Pappenheim would later become an activist who resisted, and helped others to resist, the apparatuses that interfered with women's freedoms.

2 A degree of uncontrollability is at the core of singularity, which is, then, necessary for coming to know the object or person. As noted in previous chapters, for Rosa (2020), the possibility of an experience of resonant relationships requires some surrender to the uncontrollability of the other. In this sense, which I will say more about below, resistances and obstacles are inherent to uncontrollability and singularity, and both are integral to coming to know obstacles.

3 Philosopher Michael Cholbi (2021) argues that grief and mourning are opportunities for self-knowledge. We might say that grief is perceived as an obstacle, and to work through it by way of mourning deepens self-knowledge.

4 The Borg queen indicates the presence of gender, but it is a monstrous depiction of gender in the sense that there is an absence of the capacity for affection, empathy, compassion, and intimacy or resonance with singular others. Moreover, the queen's body is not whole and is tied to and dependent on mechanical apparatuses, as are all the other Borgs. From a psychoanalytic perspective, it is worth wondering why the authors chose a female to represent the apex of the Borg collective.

5 The presence of semiosis, whether in human beings or other animals, necessarily implies resistance. Without resistance, there would be neither semiosis nor singularity.

6 This discussion may raise, for readers, a question regarding the freedom of other-than-human species. To answer this question would require a philosophical discussion of the notion of freedom, which would take me far afield. I will say, however, that the presence of the ontological rift results in the denial of freedom with regard to other species and othered human beings. Animistic epistemologies would not deny the notion of freedom of other species since freedom is linked to singularity and the actualization of living beings' potentialities. What freedom actually looks like in other species would require knowing something about them, which implies that the notion of freedom is distinct given the potentialities of each living being. I will say that anything that impedes the actualization of a creature's potentialities (singularity) is a denial of its freedom. Whether or not this is ethical is another question.

7 Rosa (2019) focuses on alienation instead of dissonance. In addition, he points out that the nihilism and violence evident in the modern age "function as a means of inducing resonance, of forcing mute or repulsive horizontal and diagonal relationships to dance in accordance with the said promise of responsiveness, of making the silent world resonate" (p. 267). Not surprisingly, the result of violence or force is alienation. I take a slightly different position, arguing that the ontological rift produces dissonant subjects in relation to other species and the Earth. This dissonance is evident in the semiotic construction and abstraction of "Nature" that shapes our objectifying and aggressive relationship toward all things deemed to fall within this abstraction.

8 From a different but related angle, Erich Fromm (1977) wrote, "The frequency and intensity of the desire to share, to give, and to sacrifice are not surprising if we consider the conditions of existence of the human species. … What is surprising is that this need could be so repressed as to make acts of selfishness the rule in industrial (and many other) societies and acts of solidarity the exception" (p. 87).

9 Available at: www.goodreads.com/quotes/841207-philosophy-is-really-homesickness-the-urge-to-be-at-home (accessed March 30, 2025).

10 We do not even live resonantly with each other; consider, for example, the political philosophy of men such as Carl Schmitt (2005), who emphasized that the central feature of political life is the friend–enemy relation—a relation characterized by dissonance.

11 Todd Mei (2017) argues that Heidegger viewed "the lack of housing as symptomatic of the failure to understand what it means to dwell" (p. 130; see also Malpas, 2006). I would add that the failure is not only about human dwelling but also about how we dwell in this place called Earth and with more-than-human species. The unhousing of other species, in other words, is a symptom of both our resistance to recognize their singularities and our dissonant dwelling in the world.

12 I mentioned in the Introduction the work of Penrose, Hameroff, and Kak (2017). Briefly, they argue that material existence manifests resonance at the quantum level, which indicates that resonance is the basis of animate, sentient semiotic activity. All living beings are semiotic. While resonance applies to material existence, there is a distinction between resonance at the quantum level and resonance in relation to semiosis and life. I suggest that the distinction is that living creatures can experience resonance, while inanimate objects cannot. Put differently, resonance at the quantum level (atoms and subatomic particles) through the levels of molecules, mitochondria, cells, bacteria, simple organisms, animals, etc., represents increasingly complex (not higher) organizations of resonance. The implication here is that resonance is a foundational feature of material existence and of all living beings. As Rosa (2019) writes, "The primal form of existence is a relationship not of alienation but of resonance" (p. 258). This does not mean there is no dissonance, but dissonance is an epiphenomenon.

13 Albert Camus (1956/1991) wrote that "rebellion, though apparently negative, since it creates nothing, is profoundly positive in that it reveals the part of man which must always be defended" (p. 19). This "part" for Camus is solidarity. Camus's view regarding rebellion has parallels to what I call political resistance and resistance in general. Political resistance and "solidarity" can be understood as the affirmation of resonance, not simply with other human beings but also with other species and the Earth. When Camus wrote, "I rebel—therefore we exist" (p. 22), he was affirming solidarity, which I would interpret as the existential resistance associated with singularity. To be a singular living being is to resist sameness and existential insignificance. And, for human beings, as social altricial animals, singularity is ensconced in solidarity with our fellow human beings. I explore this in greater depth in Chapter 6.

14 As I mentioned in Chapter 1 (footnote 23), Charles Mills (1997) defines the racial contract as a "political system, a particular power structure of formal and informal rule, socioeconomic privilege, and norms for the differential distribution of material wealth and opportunities, benefits and burdens, rights and duties" (p. 3). Mills points out that this contract is political, moral, and epistemological (p. 9), and that it "is clearly historically locatable in a series of events marking the creation of the modern world by European colonialism and the voyages of 'discovery' now increasingly called expeditions of conquest" (p. 20). Mills adds, "We live in a world which has been foundationally shaped for the past five hundred years by realities of European domination and the gradual consolidation of global white supremacy" (p. 20).

15 "Ruby Sales—Where does it hurt? On Being with Krista Tippett," *On Being*, last updated January 16, 2020. Available at: https://onbeing.org/programs/ruby-sales-where-does-it-hurt/#transcript (accessed December 4, 2025).

16 Chief Seattle speech, Suquamish Tribe. Available at https://suquamish.nsn.us/home/about-us/chief-seattle-speech/quamish Tribe (accessed January 13, 2026).

17 David Hart's (2024) book *All Things Are Full of Gods* comprises lengthy imaginary dialogues around the question of and the debates about the relation between mind and matter. In the words of Psyche, "Human beings turn for companionship to the thin, pathetic, vapid reflection of their own intelligence in their technology only

because they first seal up their ears against the living voice of the natural world, to the point that now nothing more than its fading echo is still audible to them" (p. 480). Psyche continues, "Whatever the case, after four centuries of mechanistic dogma, the inability to view the natural order as a realm of invisible sympathies and vital spiritual intelligences is very much the essence of the late modern human condition" (p. 481). In his own words, Hart affirms a kind of panpsychism or a kind of vitalism, "though by this I do not mean a theory of life as some indefinable quintessence infused from without into an otherwise mechanical nature; rather, I mean that life is itself the pervasive 'organic' logic of the material order from the first, not emerging from that order but instead creating, governing, forming, and quickening it from within. I believe, moreover, not only that mind and life are both irreducible; I believe that they are one and the same irreducibility" (pp. 6–7). Given this, it would be a mistake to ascribe "mind" to human beings alone.

Chapter 6

A Psychoanalytic Process

Facilitating Ungovernable Selves in the Anthropocene

Tyrants, totalitarian regimes, and colonizers seek to produce governable selves through propaganda, force, violence, and varied forms of intimidation. What is even better, for the sovereign classes, is the willingness of the population to accept their chains. Thomas Hobbes's political philosophy, in my view, is, at best, an unwitting attempt to convince people of the absolute necessity of a leviathan for social stability and freedom. Rouse citizens to fear anarchy by way of origin myths of chaos and violence and their imaginations will constrict, making them docile subjects of the sovereign classes. Governable selves will sacrifice freedom and desire for a social contract and the putative protection of the sovereign. Indeed, if enough fear and anxiety are evoked, mixed with large doses of propaganda, as we are seeing throughout the world today, people will cheerfully elect authoritarian leaders and their toadies. It is not simply that authoritarian leaders are Orwellian in the extreme, it is that governable citizens are unable or unwilling to distinguish between the bullshit and the lies (Frankfurt, 2005), usually accepting both as gospel from the mouths of tyrants. But it is not only docile citizens who are governable; it is also the sovereign and members of the sovereign classes—political and economic elites. They are not docile, but they are governable in the sense of being captive to their own lies and to their fear, if not terror, of losing the power, privilege, and prestige that comes with ruling over the masses. They are in bondage to the very apparatuses that secure their rule, while governable citizens are in bondage to them.

None of this is new. Plato and Aristotle were certainly aware of tyrants and understood that democracies can easily devolve into hatred and fear, wrongly executing people, such as Socrates, with the aim of keeping people from questioning what are held to be sacred truths—truths that are deemed necessary for the stability and security of the polis. What is new today is the existential threat of climate disaster. Despite numerous warnings dating back at least to Rachel Carson's (1962/2002) *Silent Spring*, the major contributors (imperialistic nations such as the United States, Russia, and China, as well as European nations) to global warming have been slow to respond or to respond at all. Opportunistic, Orwellian, nihilistic authoritarians and their

DOI: 10.4324/9781003518013-7

followers, such as Governor DeSantis[1] and the current U.S. president, vice president, and House speaker, not only publicly deny climate science and mountains of empirical data, they find ways to delete any mention of climate change from government documents, as well as remove climate funding and fire scientists.[2] While not completely successful in obtaining totalitarian control by way of bullshit and lies, they have produced sufficient compliant or governable citizens to go along, thereby undermining and obstructing efforts to stem the tide of climate disasters. Worse, their actions and policies contribute to the rise of greenhouse gas emissions.

The crises associated with climate change, which seem to give rise to authoritarians and the polycrisis they produce, raise questions about how to respond. Clearly, the emergence of authoritarians around the world is, in one sense, a response to the current and looming climate disasters and accompanying traumas. If we do not want to be docile citizens, it is not simply a question of how to respond to the crises precipitated by petty, though dangerous, tyrants. It is also a question of how to respond to the realities of climate change. More specifically, in terms of psychoanalytic theories and practices, how do we understand not only our role in relation to the needs, concerns, and sufferings of patients but also our role in terms of the larger society, as well as our ethical responsibilities and commitments to othered human beings, other species, and the Earth (Bednarek, 2024; Samuels, 2023)?

This may seem to exceed the writ of psychoanalysis, but I strongly disagree. Consider first what happened when the Nazis took over Germany. Matthias Heinrich Göring, a Nazi and relative of Hermann Göring, assumed control of the Berlin Psychoanalytic Institute. Many Jewish psychoanalysts were able to flee, but the Göring Institute and others, supported by Nazi organizations, reshaped psychotherapy and co-opted many psychiatrists and therapists (see Cocks, 1985/2018; Goggin & Goggin, 2001; Sokolowsky, 2022; Thomä & Kächele, 1994).[3] Of course, antisemitic Nazis would wish to remove Jews from any private or public position, but why co-opt psychological (in particular, psychoanalytic) training and practice? I believe it was for two interrelated reasons. First, they wanted therapies and therapists to operate as another apparatus toward producing docile citizens. Second, in my view, they feared psychoanalysis because, as Adam Phillips (1993) notes, it functions to undermine idolatries.[4] Authoritarian leaders, secular and religious, depend on creating and maintaining collective idolatries because these idols function as soporifics, dulling the mind and creating submissive followers. When you question and expose the apparatuses of idolatry and their attendant bullshit and lies, as I think Frantz Fanon understood, you facilitate the emergence of ungovernable selves who are able to recognize and choose an action toward the real social, political, and economic sources of suffering.

In this chapter, given the context of climate polycrises, which includes the rise of authoritarian types of sovereignty, I explore the relation between psychoanalysis and the notion of ungovernable selves. I begin by discussing how analytic therapies emerged at a particular Western cultural moment that had previously been dominated by commitment therapies. Juxtaposing these therapies offers clarity about the strengths and limitations of both, but, more importantly, it allows us to take note of *one* possible aim of the analytic process, namely, the facilitation of ungovernable selves who choose commitments toward the real sources of human and other-than-human suffering. From here, I move to address the notions of rebellion and revolution using Albert Camus's work, primarily because it is important to make distinctions between these concepts and ungovernable selves. I then turn to explicating the notion of ungovernable selves and its relation to psychoanalysis. The concluding section shifts to the notions of "soul" and "spirit." I do this for two reasons. First, psychoanalysis, from its inception, has had an ambivalent relation to religions, often failing to acknowledge the affinities between religion and science. Second, one way to understand the notions of soul and spirit is in terms of abstraction and resonance. That is, they are, at best, abstractions for experiences of resonant relations of belonging, which are features of ungovernable selves. In this case, then, psychoanalytic practice can consider patients' religious beliefs and experiences in light of whether they are connected to ungovernable selves' resonant experiences and relations with other human beings, other species, and the Earth.

It will be helpful here to offer a few clarifications before we begin. First, let me state clearly that what I am offering is a way of thinking about psychoanalytic therapies. I am not, in other words, implying that this applies to all patients or minimizing other important aims of psychoanalytic therapies, such as healing, reconciliation, and freedom. I remember one of my analytic supervisors saying, several decades ago, that we need to recreate therapy, given the unique needs, desires, and experiences of individual patients. So, the notion of ungovernable selves may not be applicable to some patients.

Second, I contend that therapists of any persuasion have an obligation not only to care for their patients but also to address the ills of the larger society that are sources of suffering and distress. There is, then, a political and ethical accountability in and commitment to the practice of caring for patients, society, and other species, which is especially evident in the realities associated with the climate crisis. How one manifests or lives out this political and ethical accountability varies. I add here that psychoanalytically inclined people can function and, in my view, have functioned in society to facilitate or encourage the formation of ungovernable selves.

Third, I do not think psychoanalysis is necessary for democracy (or for any iteration of sovereignty), though healthy democracies, unlike the Athenian polis, are open to Socratic questioning and analytic exploration. Socrates, in fact, strikes me as an ungovernable self residing within a democratic polis. He was seen, in part, as a rebel and, therefore, a criminal, but he denied this, proclaiming not simply his innocence but the validity of his method of seeking clarity, complexity, and truths. Though he saw himself as governable, paradoxically he was not in that he chose to accept the verdict of death, implicitly denying that the state had absolute power to demand his death without his consent. In other words, the state's verdict denied Socrates's agency and freedom, yet Socrates claimed both in his decision to accept the sentence but not the judgment. Psychoanalysis is not Socratic, but it too pursues clarity, complexity, and truth. By this, I do not mean truths are imposed but rather that clarity and truths emerge from the process of analytic exploration. In this way, analysis itself is ungovernable regardless of the polis in which it resides unless, of course, analysis is seen as the truth. When this occurs, as Phillips (1993) notes, psychoanalytically inclined persons are captive to an idol and become governable.

A Brief Retrospective on Psychoanalysis

In this section, I first locate the social and cultural origins of psychoanalysis, differentiating the psychoanalytic process from commitment or classical therapies that preceded the emergence of psychoanalysis in the late 19th century. This discussion helps identify some of the connections between these two types of therapies as well as their clear strengths and limitations and the differences between them. That is, by differentiating between classical and analytic "talking cures," we obtain a clear appreciation for the varied aims of psychoanalysis—aims that undergird the idea of ungovernable selves. Moreover, clarifying the distinctions between these approaches in light of the climate crisis reframes our understanding of commitment therapies versus psychoanalysis. From here, I identify some of the aims of analysis, as well as motivations for developing and practicing psychoanalysis. This brief overview reveals the contextual variability of psychoanalytic practice and goals and sets the stage for arguing that psychoanalysis possesses a liberative dimension. This liberative dimension is taken up in the following section on the analytic process as inviting the emergence of ungovernable selves—selves that render inoperative the apparatuses of the rift and, in so doing, recognize their existential commitment to care for other species and the Earth.

The 19th century saw the rise of secularization and individualism as well as the decline in religious authority in the West, which was captured in Nietzsche's pronouncement regarding the death of God. Nietzsche and novelists such as Dostoevsky signaled the sense of dislocation, estrangement, and disenchantment many people in the West were experiencing (Nishitani, 1982). Various other writers recognized this decline in belief in God and its consequences. Alexis de Tocqueville (2004), while lauding the possibilities of individualism in the context of democracy, was concerned that secularization would undermine the moral foundations of society, which is ironic given the immorality of slavery, ethnic cleansing, colonization, and rampant sexism perpetrated by so-called religious people in a "democratic" society. In the same century, Fyodor Dostoevsky's novels often contrasted the themes of nihilism and religious beliefs in a society where religion was in decline. Long before Freud's attacks on religion, Ludwig Feuerbach, Karl Marx, and, of course, Friedrich Nietzsche were critical of Christianity and belief in a monotheistic God. These were all features of the decline, and this decline corresponded to the rise in the authority of science to provide answers and treatments for the various physical and psychological maladies people experienced (see Gilman, 1985b; Porter, 1991a, 1991b).

Prior to this decline, commitment therapies were prevalent. For millennia, Western people who were psychologically distressed or ill, by and large, sought out religious healers, who understandably made use of religious cosmologies to understand illnesses and to develop interventions aimed at alleviating suffering so that distressed persons could regain their participation within the community and with God(s) (for a Christian perspective, see Clebsch & Jaekle, 1994). Diverse types of rituals and more informal conversational methods were used and continue to be used by religious leaders and healers.[5] Any relief from suffering through cure or the discovery of new meaning in one's suffering was understood theologically and believed to be transformative for the individual (Rieff, 2006, p. 76).

These classical therapies fall under the heading of the cure of souls tradition. In this tradition, ostensibly trained religious persons, tasked to help people suffering from soul maladies, rely on varied theological interpretive frameworks for diagnosis and intervention. Many of the more prominent healers developed methods and articulated aims of practice. For instance, Gregory the Great (1978), a sixth-century monk and later pope, wrote a book on pastoral care, outlining a method for sensitively and respectfully conversing with those who suffered from various relational and soul maladies. This method included clear theological analyses of the individual's predicament and prescriptions for interventions. Centuries later, various U.S. Protestant pastors developed and wrote extensively about their talking cures or conversational methods in the 17th and 18th centuries, hoping to teach other pastors appropriate diagnoses and effective methods for responding to soul suffering (Holifield, 1983). Clergy were expected to be experts in diagnosing the maladies of the soul, and the primary methods for dealing with these maladies were called answering methods.[6]

Some of these talking cure methods were early forerunners of cognitive-behavioral therapies, relying on *rational persuasion* to counter a person's false religious beliefs, though healers using these methods were interested in change that restored a person's relationships with the Christian God and a particular community of faith and its traditions. There were also interpretive techniques for the cognitive reframing (theologically) of persons' understanding of their illnesses with the aim of providing meanings that persons could constructively use and that would mitigate their suffering and alienation. For instance, Reverend Thomas Hooker was called to help a New England Puritan woman tormented by thoughts of guilt and damnation (Holifield, 1983, pp. 33–37).

The previous pastor's rational method had failed to convince the woman that she was not damned. Hooker reframed her alienation by suggesting to her that her suffering was an indication that God had chosen her and her suffering. Therefore, her suffering had nothing to do with her guilt or being damned. Reframing the meaning of her suffering from guilt and damnation to God's presence amid suffering led to a change in the woman's attitude and mood. That is, this young woman, according to Hooker, was no longer tormented by guilt and was able to find meaning in her suffering. Regardless of what one thinks about the theology manifested in this reframing, it was, to Hooker and the woman's family, very helpful in relieving her guilt and fear of damnation and restoring her to her family and the community of faith. Whether this was a cure or simply symptom relief is difficult to know, but persuasive conversational approaches are not uncommon in some current secular therapies (e.g., Frank, 1961). However, the secular interpretive frameworks used to reframe the meanings of a patient's suffering today are not theological, and the therapeutic aims are decidedly different.

Phillip Rieff (2006) placed these talking cures, as well as other rituals of care, under the heading of commitment therapies—a term I agree with, though with some qualifications. Healers were recognized by the community as possessing the knowledge, skills, and authority to care for people. A healer's aims were not simply to reduce or eliminate suffering. To be sure, to cure or remove suffering was good in itself, but the ultimate aim was to restore the person to the community and to God. Suffering, which could be the result of the individual's sin, possession, or unexplained natural suffering (natural evil), often led to emotional experiences of alienation from the community and from God—undermining persons' commitments to both. This is decidedly different from Rieff's (2006) claim that classical caregivers sought to "commit the patient to the symbol system of the community, as best he can and by whatever techniques sanctioned (e.g., ritual or dialectical, magical or rational)" (p. 68). Rieff seems to think that the patient was not using the symbol system of the community, which in many cases does not make sense since there were no other symbol systems for the individual to commit to or to use to understand their malady. More importantly, someone who suffers does not necessarily suffer as a result of having failed to commit to a symbol system. The sufferer is also not *initiated* into a community and its beliefs system. Instead, the symbol system they have long unwittingly internalized is used to derive meaning from, solace for, and a cure for their malady. To be sure, sufferers may feel alienated from others of the community and from God, but the experience of alienation is understood within the context of the religious symbol system they have long internalized and used.

In my view, "classical" soul therapists or commitment therapists used (and use) the symbol system to assist in understanding and responding to suffering and, if possible, restore—not initiate—persons' capacities to commit to family and community (not to commit to the symbol system). When individuals, through no fault of their own, could not be restored to family and community, the question became how the community could contain and sustain them—at least those persons who had not, through their own sin, caused the alienation (Clebsch & Jaekle, 1994).[7] In short, these religious talking cures and other rituals were not necessarily initiation rituals as Rieff argued.[8]

It is important to note that these classical therapies addressed two basic sources of suffering. In instances when suffering was the direct result of human actions, classical therapists sought to have the person take accountability for (confess) their actions and perform acts of restitution (penance) to repair their relationship with people in the community and God (Ellenberger, 1970, pp. 22–25). Again, this did not mean helping the person to commit to a symbol system; rather, the aim was to restore their commitments and relationships with others and with God. On those occasions when the suffering was not the result of an individual's decisions and actions, the challenge was to cure them or, when that was not possible, provide some solace, meaning, or purpose within the context of a sustaining community. In both cases, as well as those when the meaninglessness of suffering threatened, the healer and community ideally provided sustenance, comfort, and consolation. Again, these were collective actions of commitment that aimed to provide emotional support in the face of physical and psychological suffering that was understood to be alienating.

It is significant to note that my generalizations are taken from Western religious traditions, mainly Christian. I now, with some trepidation, expand the focus to Indigenous peoples, though there are important differences. Andrew Solomon (2015) writes of his experiences of various Indigenous rituals used to cure depression, though Indigenous peoples would not use the term "depression." Solomon, who underwent some of the therapies, found some temporary relief but, in the end, he continued to suffer from depression. Perhaps he was, understandably, looking for a cure, but it is not entirely clear that the Indigenous rituals were manifestly directed at a cure. They could have been aimed at providing meaning and communal sustenance for the sufferer. In his cross-cultural exploration of schizophrenia, Louis Sass (1992), relying on the works of various anthropologists, notes that Indigenous communities, while having a lower rate of schizophrenia than industrialized societies, possess different ways of understanding and responding to psychological maladies. The aims were and are to relieve suffering and to provide the sufferer the solace and support of the community when suffering cannot be relieved.

This is similar to classical Western religious therapies that use the symbol system to understand and respond to maladies. One significant difference, generally speaking, is the absence of the ontological rift in Indigenous apparatuses. This means that Indigenous interpretations of maladies do not contain the philosophical or theological premise of ontological alienation and the attendant existential dissonance (with resonance being an epiphenomenon of living beings). In other words, religious ideas of original sin and salvation as well as philosophical ideas of human existential homelessness are not part of Indigenous mythologies. What this means in terms of approaches to care would take more time and space than I have here. However, a general hypothesis is that Indigenous talking cures or commitment therapies are aimed at reestablishing resonant relations between the person, the community, the spirits, and the so-called natural world and its diverse inhabitants. The underlying premise of these therapies is that dissonance is an epiphenomenon of resonance,[9] while the underlying premise of classical Western religious therapies is that resonance is an epiphenomenon of dissonance.

These classical therapies, Indigenous and otherwise, also possess several other underlying and interrelated premises. First, the self/soul or person is inextricably yoked to community, and the community is a subset of the larger society. A self, if you will, has its being and life in community (cf. Macmurray, 1949/1993). To be exiled is to die or to experience horrendous suffering, as seen in various depictions of hell. This idea that persons have their origin and development in community is an ancient theological (and philosophical) idea that continues to be an integral part of Abrahamic religious anthropologies (e.g., Niebuhr, 1989; Volf, 1996; Zizioulas, 1985, 2006). In these theological anthropologies, community is not simply for the sake of survival but for living a meaningful, resonant life with others in the community. A difference between these and Indigenous anthropologies is that Indigenous notions of a polis or community includes other species and the land. This difference is quite significant with regard to the climate crisis and the presence of the ontological rift between Western subjects and other species. It is significant because these classical therapies do not attend to the dissonance between Western human beings and what we call "nature." I add here that classical Western therapies share this issue with analytic therapies. As I have argued throughout this book, analytic therapies, while having many benefits, are in the main not concerned with patients' relationships with and experiences of other species. I will return to this below.

We can shift from these religious and philosophical premises regarding self and community to a modern therapeutic angle to affirm the premise regarding self and community. Judith Herman (1992) points out that traumatic events "destroy the sustaining bonds between the individual and community" (p. 214). Though using different language, she argues that the third stage of good-enough therapy is to help patients reconnect and regain a sense of commonality (p. 236). The major difference is that analytic therapists are not representatives of a particular community and, therefore, do not aim patients toward this or that community. Analytic therapists acknowledge the importance of intimate relationships (e.g., family, friends, community) and may evaluate them by relying on psychoanalytic theories. However, unlike classical therapies, therapists do not aim the person toward a particular community and tradition.

A second key and related premise of classical therapies is the notion of an encumbered self. Human beings make commitments and depend on the commitments and promises of others for their survival, their resonant experiences with others, as well as a sense of purpose and meaning. This encumbered self is not to be understood as an individual weighed down by or onerously confined to communal and family obligations. To be sure, commitments can be difficult, even burdensome, but, for classical therapists, having an encumbered self also means living a meaningful and purposeful life through commitments to and with others. The religious notion of covenant and its centrality in relation to God and community signify the importance of encumbrance in human life. Indigenous peoples also stress the importance of fidelity to the community and its practices, though they would extend fidelity to other species as kin. Fidelity or keeping the covenant means life—physically and psychosocially. Conversely, breaking covenantal relationships has been represented by stories of illness, death, and foreign bondage and oppression.[10]

While this may seem like a strictly religious idea, there are numerous examples of nonreligious people choosing to be encumbered and, in doing so, experiencing communion, meaning, and purpose (see Kegley, 1980; Royce, 1908/2018). Good-enough parents, for instance, are encumbered by voluntarily being obliged to meet the numerous and intense demands of their baby. Without their commitment to the infant's physical and psychological well-being, there is little hope that the baby will survive, let alone thrive. People who get married decide to encumber themselves with each other. A husband is encumbered by his wife's illness, finding the alternative untenable, yet also finding meaning and purpose in the shared struggle. A therapist, working with a borderline patient who is suicidal, chooses to be encumbered so that there are the hope and possibility not only of the patient's survival but also of a better life. Frantz Fanon clearly took up the weighty responsibility for and commitment to treating patients in Algeria and working with others to free Algeria from the French colonial yoke—a yoke that was damaging to the psychological and relational health of many Algerians. An unencumbered self, in short, is inconceivable for classical therapists because their commitments to family, community, and God(s) are foundational to individual and collective survival and life. I think that the idea of an unencumbered self, unbound by any obligations and commitments, is not an aim of analytic therapies. Of course, patients who are encumbered by past traumas or encumbered in the present by oppressive relationships do need a process that facilitates not only their freedom but healthier decisions regarding choosing encumbrances that are resonantly life-giving.

This leads to the third related premise, and that is the notion of freedom. For classical therapists, freedom is not to be understood as freedom *from* the exigencies of living in family and community, at least not entirely. Freedom is experienced in relation to other members of the community and in their mutual commitments (Macmurray, 1949/1993, pp. 31 ff). Encumbered selves are free, by voluntarily committing themselves to others, to living a life in common—free to commit, to take up their encumbrances, which, in turn, provides individuals with meaning and purpose. The paradox is that individuals are free if they voluntarily encumber themselves with others. For Indigenous peoples, this encumbrance and freedom include responsibilities to other species and the Earth.

For classical therapists, then, freedom is not an absolute or individualistic, as if human beings could live without constraints, encumbrances, limits, and so on. Freedom is conditional; that is, freedom is conditioned by one's decisions to accept the encumbrances and the consequences that attend one's obligations to others. Good-enough parents voluntarily encumber themselves and, in so doing, make possible their child's development as a human being who, in time, will voluntarily commit to their family and community. The child, in other words, will learn that freedom involves the experience of, and decision to live, a life in common with particular others. Being free from all commitments, from a classical therapy point of view, would lead to alienation, dissonance, emptiness, meaninglessness, and purposelessness. More to the point, an unencumbered self, for classical therapists, if conceivable, is not free but is a slave to their own fears, desires, and wishes.

Of course, classical caregivers recognize that there is also the idea of freedom *from* commitments or situations that are unjust, oppressive, or destructive. For instance, Israelites were liberated from the Pharaoh and from Babylonian exile. Yet *freedom from* did not mean an escape from their obligations to their people or to God. Being free from unjust relations allowed the Israelites to be more fully committed to the well-being of their people and to the community's covenant with God. Ideally, becoming free is a return to an encumbered self who, after being freed from the chains of oppression, freely chooses to commit to the flourishing of their people. Classical therapies, then, are ideally aimed at freeing the individual from unnecessary suffering that causes alienation and diminishes their ability to engage in committed, meaningful relationships with others.

I will briefly jump ahead here and suggest that analytic therapies clearly have a liberative feature as well—a liberative feature that is relational. Fanon believed that the aims of analytic therapy included raising to consciousness the real social and political sources of the patient's suffering so the patient can choose an action toward those sources. These aims actually parallel Fanon's life and political development. He became increasingly aware of social, political, and cultural sources of racism and colonization that undermined the psychosocial well-being and freedom of Algerians. He repeatedly and freely chose to encumber himself in the cause of solidarity and liberation. His freedom from the psychological fetters of colonial racism gave rise to his freely choosing to help patients in Algeria who were suffering under brutal French colonizers, as well as choosing to take an active political role in national liberation (Gordon, 2015).

So far, I have been focusing primarily on individuals and their sufferings in terms of the aims and premises of classical therapies. It is necessary to point out that healers also have other concerns and aims, which could undermine care and concern for some sufferers. As representatives of their community and its tradition, healers and religious leaders are also concerned with the security and stability of the community and its tradition, which means that individuals, at worst, are sacrificed for these aims.[11] Identified leaders and healers, in other words, are responsible not only to the "patient" but also to the community of faith and to God. For instance, St. Paul, in his pastoral letters, at times counseled people within the faith community to seek reparation with certain community members and, if this did not work, to eject the persons from the community. Also, we read in Hebrew and Christian scriptures about the marginalization of individuals who suffered from leprosy because of fears they would contaminate others and undermine the health and well-being of the community. And, for centuries, those who suffered from what we now know were most likely mental illnesses were exiled or ostracized (see Foucault, 1965, 1987; Gilman, 1985a, 1991; Porter, 1991a, 1991b; Szasz, 1977, 1987). In these instances, the survival and putative well-being of the community or society took precedence over the individual.[12]

Since I mentioned this limitation, let me identify some advantages and disadvantages of classical commitment therapies. First, by relying on the community's symbol system and rituals, individual sufferers can find meaning and purpose in their struggles within the context of committed relationships, ideally relieving them of the psychological suffering that comes from meaninglessness. Second, classical therapists seek to return persons to the community, where they can experience communion, meaning, and purpose in their commitments to others. Third, when suffering is not going to be relieved or cured, classical therapists, along with the community, ideally help sustain the patient. That is, commitment therapies reinforce the commitment of the community to the sufferer, which can help attenuate the suffering that comes from alienation. Fourth, the wisdom of classical therapies is the existential, paradoxical recognition that encumbered selves, in good-enough relationships and communities, experience freedom in their voluntarily encumbering themselves with others in the family and community.

There are, of course, important disadvantages to commitment therapies, not the least of which I mentioned above regarding apparatuses of the ontological rift. Rieff (2006) was partially correct when he stated that commitment therapies were authoritarian (and possibly idolatrous, à la Adam Phillips, 1993). Certainly, they could be authoritarian whenever meaning and commitment were imposed on those who suffer. Worse, there was authoritarianism in "therapies" that led to exile or death for those who were perceived to threaten the community and society (burning heretics is a counter-initiation ritual). It is a mistake, however, to suggest that commitment therapies were (or are), in and of themselves, authoritarian or, by contrast, that analytic therapies cannot be authoritarian. Any cursory reading of clinical cases in psychoanalytic history reveals, at times, the presence of authoritarianism in interpretations being imposed onto patients. Still, this does not mean that psychoanalytic therapies are, in and of themselves, authoritarian.

Another disadvantage of commitment therapies is the often-unquestioned interpretive religious framework used to construct diagnoses and interventions. The theological worldview is often taken to be unquestionable, and this may attend the expectation of an unquestioned acceptance of the community, its traditions, and its way of being in the world. So, in some situations, the individual sufferer and healer may not see that their suffering results from the community and its tradition. A case from the 17th century illustrates this view (Holifield, 1983). The patient was a young woman whose parents married her off to an older gentleman. Soon after giving birth, this woman began to have visions and thoughts of damnation. It was inconceivable to the parents and the pastor, who was using a talking cure, that this woman's suffering may have been related to her having been forced to marry. In other words, the political and religious apparatuses of patriarchy resulted in this woman's suffering. It was seemingly impossible to consider alternative interpretations or interpretations that contradicted the dominant religious and political views about young women and marriage.

The disadvantage of not questioning one's religious (or secular) worldviews is especially problematic today given the destructive realities of the climate crisis and how Western philosophical, religious, and scientific apparatuses are implicated. Many of us, in a sense, are like the woman suffering in the 17th century who was trapped in a symbol system that was unquestioned, leaving her without a remedy. For subjects of the rift who do not or will not question Western apparatuses, possible remedies for our dissonance will elude us. Max Horkheimer presciently commented, "The future of humanity depends on the existence today of the critical attitude" (as cited in Wolin, 2016, p. 231), and this critical attitude entails questioning our theories and practices.

Another related disadvantage of classical therapies is the tendency to encumber individuals who do not need to be encumbered because they are already overburdened or those whose suffering is directly related to their current distorted encumbrances. A painful example of this is a clergyperson telling an abused woman she must work harder at being a good wife—a view "rationally" legitimated by the religious tradition. In this situation, the woman's encumbrance and freedom signify a distortion. She is involuntarily subjected to the abusive husband, largely out of fear and partially out of her belief that her very sense of self depends primarily on this relationship rather than on a sense of self that is situated within and dependent on a wide range of communal relationships of mutual commitment, respect, and affection. Moreover, the husband's (and clergyperson's) view of encumbrance is distorted because his freedom is dependent on the subjugation of his wife instead of arising out of mutual and voluntary encumbrances associated with the religious notion of covenant.

A final limitation of classical therapies is that their helpfulness can be narrow in their reach or effectiveness. As pointed out above, classical therapies emerge out of a particular community and its symbol systems. Generally, those who do not accept the community's common purpose, commitments, and symbol systems are not subject to or subjects of the rituals for healing or cure. An atheist who is dying of cancer is likely not going to find it meaningful or helpful to have a clergyperson offer prayer and anointing. Commitment or classical therapies, then, can be parochial and, in secular societies, not available or helpful to the wider public or to nonbelievers.

In brief, the cure of souls tradition involves commitment therapies that include talking cures. These therapies rely on theological interpretive frameworks for diagnoses, interventions, and therapeutic aims. Analysis is part of the diagnosis, but this entails listening for the hidden movements of the soul or occult aspects and sources of evil and sin. The aims of commitment therapies include (1) healing or mitigating the individual's malady; (2) providing meaning and purpose amid the individual's suffering; (3) repairing the individual's relationship with God; (4) restoring the person's commitment to the community of faith; (5) sustaining and deepening the person's experiences of communion with God and others; (6) providing communal support and solace for those whose suffering is incurable; and (7) ensuring the security, stability, and well-being of the community. Unfortunately, Western religious commitment therapies, for the most part, replicated and still replicate the apparatuses of the rift, leaving unexamined the dissonant relations between human beings and other species.

The rise of secularization and the decline of religion created spaces for new types of therapies to emerge. These therapies were not tied to religious apparatuses of any kind, though, as noted in Chapter 1, this does not mean that the new analytic therapies were or are entirely free of the apparatuses of the rift. I have already mentioned Bertha Pappenheim's role in Freud's developing an analytic talking cure. Unlike religious commitment therapies, analytic therapies depended on the apparatuses of Western science. When Freud was confronted by James Putnam regarding the absence of moralistic aims in psychoanalysis (Hale, 1971), Freud, in my view, balked at any mention of morality because it smelled of commitment therapies and their tendency to frame suffering theologically and morally. For Freud, the aims of psychoanalytic therapies are to raise to consciousness what is unconscious and to facilitate the symptom to speak. He was not interested in representing a particular community, religious or secular. However, Freud saw himself as steeped in science and its methodologies, which was another reason he gently dismissed Putnam's concerns. Laura Sokolowsky (2022) adds, "What Freud saw was the danger of transforming psychoanalysis into a healing technique, of adapting it to social [and religious] norms" (p. 1). One could argue, though, that Freud wished people to adopt, and perhaps commit to, the norms of science.

Sokolowsky (2022) also argues that

> ultimately, the only power of psychoanalysis is the power of speech, not the power of the psychoanalyst. Psychoanalysis certainly has therapeutic effects, but it does not proceed from the desire to cure; it does not aim at relief of psychic suffering, at access to greater liberty or at the strengthening of individual autonomy.
>
> (p. 2)

The power of speech is reflected in facilitating the patient's symptom to speak or to raise to consciousness what is unconscious. One might say that the power of the analytic therapist is not the power to cure but the power to facilitate the lessening of suffering or greater liberty by way of analytic listening, which entails the use of the therapist's subjectivity to expand patients' self-awareness and insight. The benefits of this process are, ideally, some relief from suffering (to live a normally miserable life) and greater freedom and differentiation, or what Sokolowsky calls autonomy.

I suggest that Sokolowsky is only partially correct. Freud repeatedly mentioned the therapeutic aims of analysis. And as for cure, Freud commented to Jung that psychoanalysis "is in essence a cure through love" (McGuire, 1974, pp. 12–13). All of this suggests that Freud and other analysts were not simply scientists observing and understanding human behavior and psyches. They were motivated to relieve suffering and, if possible, "cure" patients. Moreover, to link a cure with love is to affirm the wedding of analytic exploration with care or love. Of course, classical therapists could make the same claim about love and its relation to cure, but there is something distinctive and new in analytic therapies. Cure and love in commitment therapies are tied to a particular community and its traditions. Analytic love is not tied to a particular community or tradition (except the tradition of analytic theories and practices). Indeed, in my view, this *analytic love is a kind of anarchic care.* There is no sovereign, human or otherwise (e.g., tradition), to accept or submit and commit to, which may be one reason why the Nazis took over analytic therapies with the aim of securing the sovereignty of their rule.

A cure through love may conjure up all kinds of warm feelings and thoughts. However, analytic love is unsettling, and not simply because it is anarchic. Sokolowsky (2022) argues that psychoanalysis is threatened when it adapts and assimilates—when it is no longer unsettling (p. 3; see also Cushman, 1995). Unsettling may be understood in several ways. First, there is a Socratic parallel with psychoanalysis. Socrates was a gadfly who unsettled the friends he loved with his questions and comments, though, apparently, he also unsettled the democratic rulers. Put differently, although Socrates pursued wisdom, he recognized that one may obtain a glimpse of wisdom but never attain it. The pursuit of wisdom is a way of life, and wisdom is dangerous or unsettling when it exposes social, political, cultural, and religious illusions—illusions taken to be truths.

Wisdom, however, is not an analytic aim, but analysis is unsettling in that it is a process that aims to bring to consciousness what is unconscious, which is a lifelong activity. What is unconscious is usually there because we fear or detest something about ourselves. We believe it to be intolerable or shameful. To be sure, the unconscious can be a source of creativity, but more often than not it is a source of that which we fear about ourselves and others. Of course, this is not a once in-a-lifetime fear. Rather, like wisdom, raising what is unconscious to consciousness is a lifelong, unsettling pursuit. Second, to pursue mindfulness in a society founded on commitment therapies that are linked to the apparatuses of the rift is dangerous to the status quo of human control, dominion, and superiority. Put another way, it is profoundly unsettling to dissonant subjects to confront their existential insignificance and the impermanence of their cherished significations.

A third difference is that, unlike commitment therapies, analysis does not offer certainty about what meanings, beliefs, values, and so on, to live out of, which can also be unsettling. There is no magisterium or Bible to commit to. There is no God to secure meaning and value. Naturally, analysis, which depends on analytic theories, can provide some clarity about the conscious and unconscious meanings and values that shape one's dispositions and behaviors, but it possesses no dogmas to commit to. This can be unsettling, especially when people in distress want clear answers and want them now. As Phillips (1993) notes, psychoanalysis is a cure for idolatry and only becomes a problem when it itself is idolatrous. Fourth and relatedly, analytic therapies can be unsettling because there is no authority to appeal to in order to secure truth claims. Socrates had Delphi, and commitment therapies have God, religious leaders, and tradition. The authority in psychoanalysis, if you will, is the analytic process, together with the skills of the therapist to facilitate the process of self-awareness. There is no final or ultimate authority to which to appeal for truth. This is not to say there are no truths, but psychoanalysis offers ambiguity and complexity, not Truth. It is likely that Freud, in rejecting God and religious commitment therapies, made science the ultimate authority, but science, despite its moments of autocracy, is ideally neither dogmatic nor preoccupied with absolute truths. Of course, Freud and other psychoanalysts could and can be quite dogmatic and even authoritarian when it comes to a particular theory or school of thought. Nevertheless, believing psychoanalysis to be a science means, in the main, being open to changing theories when previously accepted theories are shown to be incorrect or in need of further elaboration. What can be unsettling, then, is a future of change, yet this can create a space for something new—something not dependent on the authority of tradition, Truth, and so on.

This openness to change is evident in the varied analytic theories and aims that emerged during and after Freud. Alfred Adler was one of the first to offer different psychosocial notions, such as organ inferiority and children's need for affection—a need not tied to sexual instinct (Breger, 2000, p. 199; see also Gay, 1988, pp. 221–224). Carl Jung, Sándor Ferenczi, Theodor Reik, Otto Rank, Melanie Klein, W. R. D. Fairbairn, Ian Suttie, Donald Winnicott, and others followed with their own ideas about the therapeutic process and psychosocial development. Psychoanalysis splintered into schools, such as object relations, Kleinian, Lacanian, and relational. If anything, the history of psychoanalysis is a testament to the lack of an overarching authority, leaving spaces for the emergence of rebels and revolutionaries. One rebel, Otto Rank, wrote in 1932 that "psychoanalysis has become as conservative as it appeared revolutionary; for its founder is a rebellious son who defends paternal authority, a revolutionary who, from fear of his own rebellious son-ego, took refuge in the security of the father role" (as quoted in Breger, 2000, pp. 323–324).

Freud and his theories are not sacrosanct among his followers, which is not to say his work does not influence people. Rather, psychoanalysis has no ultimate authority to which to appeal for the truth or answers. Perhaps it is better to say that psychoanalysis, despite its roots in Western philosophies of the rift, is distinct from classical therapies because it contributes to the emergence of ungovernable selves.

Governable Selves, Rebels, and Revolutionaries

Before turning to the notion of ungovernable selves, I address what I mean by governable selves, as well as rebels and revolutionaries. One reason for doing so is that ungovernable selves may be attributed to those who are considered rebels and revolutionaries, yet that is not always the case. In other words, rebels and revolutionaries may themselves be governable selves. Moreover, ungovernable selves do not have to be rebels or revolutionaries. They may simply, through their way of living, render inoperative the apparatuses of the rift, like Ruby Sales's parents.

Governable selves emerge in relation to apparatuses that affirm the necessity of sovereignty for group stability and security. Governable selves believe that submitting to the sovereign and sovereign classes ensures a measure of freedom for citizens to pursue their aims. Jean-Jacques Rousseau had a different, more democratic political theory, but his idea of the general will is just another version of Thomas Hobbes's leviathan or sovereignty. Western political philosophies and theologies produce the idea that sovereignty, in one form or another, is existentially necessary for the organization of social life and, since the Enlightenment, freedom. This means that the apparatuses of government, whether in the form of democracy or monarchy, seek to produce governable selves—selves that are willing to submit to the authority of the sovereign and sovereign classes. The social contract, in "civilized" nations, requires submission to being ruled.

This is a strong claim, and yet there is evidence of anxious preoccupation regarding sovereignty in Western political philosophy. Consider what has happened to the term "anarchy" in the West. Anarchy is a political concept derived from the Greek *anarkhia* or *anarkhos*, which has a long, convoluted, complex, and contested history. Alex Prichard (2022) states that anarchy, as far back as Plato, became associated with "people misunderstanding, or acting contrary to their naturally pre-ordained status in society" (p. 5). The "naturally pre-ordained status" is that of an individual subject's obedience to the commands of a sovereign—divine or human. Hobbes, who depended on sovereign classes for his privileges, associated anarchy with chaos and violence, and that association largely continues to this day. Indeed, the very idea of the absence of sovereignty for organizing political life must have nauseated and terrified Hobbes, who railed against the idea of anarchy (political belonging without a ruler) (Ryan, 2012, p. 297). To be fair, it was not simply the loss of the privileges of the sovereign classes that struck fear into Hobbes and others. Rather, it was that sovereignty was bound up with the very notion of the self and freedom, which were believed to be dependent on a stable society. And a stable society can only exist, in this line of thinking, by virtue of a sovereign—divine or secular. The responses to the notion of anarchy were (and are) ridicule, disparagement, dismissal, and sometimes imprisonment, violence, and state-sanctioned killing. A governable self accepts sovereignty because an ungovernable self represents the terrifying possibility of chaos and loss of freedom.

This fear and this anxiety regarding the absence of sovereignty is not simply a feature of Western political philosophies. The Abrahamic traditions are also bound to an ontology of sovereignty. It is believed that creation itself is dependent on a sovereign God. Moreover, social-religious communities and societies are dependent on the sovereignty of God for stability and security. The Judeo-Christian scriptures reveal a consistent thread of God's sovereignty, which becomes, not surprisingly, wedded to human sovereignty. The elders of Israel, for instance, pleaded to Samuel to "appoint for us a king to govern us like other nations" (1Sam 8:5), which God reluctantly did after issuing a warning about human sovereigns—a warning consistently ignored then and now. The theme of divine and human sovereignty is repeated throughout scripture and is today celebrated in weekly if not daily liturgies. This serves to reinforce the idea that sovereignty is an ontological fact but also that believers, if they wish to remain in good graces, must be governable selves.

Whether rooted in philosophies or theologies, sovereignty and governable selves are, more often than not, followed by a long trail of rebels and revolutionaries throughout "civilized" history. Consider the founding Abrahamic origin myth wherein Eve and Adam break God's one command. In this and numerous other passages, Satan rebels against or seeks to undermine the sovereign God's authority or rule. The Israelites are repeatedly called a stiff-necked people, ostensibly by God (Exod 32:9, 33:3, 5; 34:9; Deut 9:6, 13; 10:16), suggesting that some of them were not entirely happy being governable selves. Milton's *Paradise Lost* is a tale of rebellion against a sovereign God. Throughout Western history we see a trail of rebels and revolutionaries. It is safe to say that, when sovereignty appeared, when a man (it was usually a man) was a ruler over other people, a rebel or revolutionary was not far behind.

Albert Camus's (1956/1991) book *The Rebel* can further this discussion on sovereignty and governable selves. "Man's solidarity," Camus claimed, "is founded upon rebellion, and rebellion, in its turn can only find justification in this solidarity" (p. 22). Camus stressed, "I rebel, therefore *we* exist" (p. 22, emphasis mine). The rebel says no to the ruler and ruling classes. In this negation, for Camus, is an affirmation of solidarity and dignity (p. 105) in the face of the apparatuses of sovereignty that undermine both. Interestingly, this spirit of rebellion "can exist only in a society where a theoretical equality conceals great factual inequalities. The problem of rebellion, therefore, has no meaning except within our own Western society" (p. 20). Camus later links this to Christianity and the notion of metaphysical rebellion against the Abrahamic God (p. 28). Metaphysical rebellion, for Camus, affirms human solidarity (I rebel; therefore we exist).[13] Whether writing about a metaphysical rebellion or a rebellion in society (or both), Camus, in my view, is noting that Western societies have always produced rebels and revolutionaries because sovereignty creates inequalities, which in turn undermine solidarity and dignity. Hence, to rebel, for Camus, affirms both.

Camus, in my view, was arguing that rebellion and revolution are inevitable realities of Western societies that found societal stability on the belief that sovereignty, divine or otherwise, is necessary and that governable selves are essential for societal existence. We obtain a glimpse of this in the Jewish scriptures when, as the story goes, God tells Samuel to inform the elders what will happen if God grants them a human sovereign.

> These will be the ways of the king who will reign over you: he will take your sons and appoint them to his chariots and be his horsemen, and to run before his chariots; and he will appoint for himself commanders of thousands and commanders of fifties, and some to plow his ground and to reap his harvest, and to make his implements of war and equipment of his chariots. He will take your daughters to be perfumers and cooks and bakers. He will take the best of your fields and vineyards and olive orchards and give them to his courtiers. He will take one-tenth of your grain and of your vineyards and give it to his officers and his courtiers. He will take your male and female slaves, and the best of your cattle and donkeys and put them to his work. He will take one-tenth of your flocks, and you shall be his slaves.
>
> (1 Sam 8:11–17 NRSVUE)

As Samuel predicted, a human leviathan and his sovereign classes possess the power to force people to be governable selves. The result is the perennial emergence of rebels and revolutionaries. Camus (1956/1991) added to this: "A religion that executes its obsolete sovereign must now establish the power of a new sovereign" (p. 121), which entails a human being and attendant sovereign classes. Of course, the elders were not killing God, but they *were* usurping the long-held tradition that only God could be sovereign. And here is the repetitive reality of the Western history of rebellion and revolution. Once a sovereign and sovereign classes are overthrown, a new sovereign takes their place, leaving only governable selves. "Every revolutionary," Camus wrote, "ends by becoming either an oppressor or a heretic" (p. 248).

Camus affirmed human solidarity and dignity in acts of rebellion and revolution. However, "Rebellion," he stated, "cut off from its origins and cynically travestied, oscillates, on all levels, between sacrifice and murder" ((1956/1991, p. 280). There is, then, a fine line between rebellion that affirms dignity and solidarity and rebellion that leads to murder and sacrifice. Indeed, they may, at times, exist hand in hand. Rebels may affirm the solidarity and dignity of their comrades while justifying and legitimating the murder or sacrifice of their opponents. Those subjugated may return the favor down the road, and then we have a vicious cycle of rebellion and revolution down the ages. Every form of sovereignty gives birth to rebels and revolutionaries who affirm solidarity and dignity yet replicate sovereignty and attendant governable selves.

I need to be clear that governable selves are not simply citizens who, willingly or unwillingly, submit to the sovereign and sovereign classes. The sovereign and sovereign classes are also governable selves in at least two ways. First, they accept the belief that sovereignty is existentially necessary for social stability and security. Naturally, this is self-serving in that they obtain the privileges, prestige, and power accorded to their rank. Nevertheless, they, like the citizenry, accept the illusion that there is no alternative to sovereignty. A second facet of sovereign classes being governable selves is that "they become slaves of their own domination" (Camus, 1956/1991, p. 176). Without their domination, they realize their sense of self is founded on an abyss of emptiness or fear of "chaos."

Camus's comment is also applicable to the climate crisis and the ontological rift. Dissonant subjects of the rift believe they have dominion over other species. While other species are considered completely governable (subjugated), humans are not unless they are depersonalized. The climate crisis reveals that dissonant subjects have become slaves of their domination of other species and the Earth. In other words, in subjecting other species to human dominion, we lose any freedom to encumber ourselves to commit to caring for other species and the Earth. In so doing, we retain our belief in human dominion while tragically undermining the very habitat on which all life depends.

Camus argued that rebellion and revolution are perennial features of Western society. From my perspective, this is because Western societies and their attendant governable selves are wedded to the idea that sovereignty is existentially or ontologically necessary for creation and social order. What I am claiming is that embedded in this philosophical and theological tradition is the tendency toward social (and nature) alienation and dissonance linked to rigid attachment to sovereignty and apparatuses of the ontological rift. A sovereign and sovereign classes emerge and survive by virtue of political violence and intimidation, which are legitimated by the apparatuses of the state. This leads to great social inequalities and indignities, which motivate rebels and revolutionaries to affirm solidarity and dignity. Yet, what remains is another iteration of sovereignty and governable selves.

In this book, I have been thinking with Agamben and Indigenous philosophies. Indigenous peoples, who have apparatuses of organizing their communities, have a dearth of rebels and revolutionaries. They do not organize their societies through notions of sovereignty and apparatuses of the ontological rift. In general, they do not see themselves as sovereign over other species or the Earth or over other members of the group. While there is, of course, conflict, dissonance, and violence, Indigenous peoples pursue resonant relations among themselves and the other species that inhabit the land. In this sense, they are ungovernable selves in relation to the ungovernable selves of other species. In other words, there is, at the basis of their epistemologies that order society, uncontrollability, which is, for Rosa (2020), a key factor of resonant relations and experiences. Other species can be engaged, understood, and killed, but they are uncontrollable because they possess personhood or singularity. In terms of organizing intergroup relations, a central principle is resonant relations; dissonance is an epiphenomenon. By this I mean that resonant relations are the guiding principle in organizing society and relating to other species and the Earth, which means that these relations produce ungovernable selves.

The history of U.S. colonization illustrates these points. Western European colonizers, and later U.S. citizens, sought to control, expel, subjugate, or assimilate the Indigenous peoples they encountered. Of course, colonizers were successful in taking Indigenous peoples' lands and nearly successful in eliminating their cultures. Despite compulsory removal from sacred lands, forcing Indigenous children into Eurocentric education, denial of cultural and religious practices, and Christian evangelization, many Indigenous peoples held on to the "old" ways. In my view, they were ungovernable for several reasons. First, in their encounter with Westerners, Indigenous selves would not initially understand Western subjects' desire for control and domination of the mute and dumb land (and Indigenous people). Nor would they initially grasp that they were being constructed in terms of Western apparatuses of sovereignty. That is, Westerners would have constructed Indigenous peoples as subject to their sovereignty (we still do) and, thus, subject to being controlled, subjugated, assimilated, or eliminated. To Westerners, Indigenous peoples were baffling, in large part, because they were not governable. That is, Indigenous peoples had other ways of organizing society that did not depend on the apparatuses of sovereignty and its inherent violence.

Second, Indigenous peoples resisted being assimilated into Western ways of thinking and acting, especially toward what Westerners called nature. In this way, they rendered inoperative Western apparatuses of sovereignty—secular or divine—simply by adhering to their ways of thinking and behaving vis-à-vis other species and the land. Third, in holding on to their customs, they affirmed that (1) selves, whether human or other-than-human, and community are not dependent on apparatuses of sovereignty and (2) resonance is the central philosophical and religious/spiritual organizing principle in their relations with each other and the Earth. Indigenous peoples were not revolutionaries or rebels, though many of them rebelled against Western attempts to subjugate them. They did not, however, attempt to replace one form of sovereignty for another, which is another indicator that they were ungovernable selves. More positively, as ungovernable selves, Indigenous peoples simply sought to live resonantly with each other and with others who resided with them on sacred lands.

This foray into distinguishing between rebels, revolutionaries, and governable selves is important to further differentiate between commitment and analytic therapies. Before the emergence of psychoanalytic therapies, Western commitment therapies, which had beneficial results in some cases, were premised on governable selves who, consciously or unconsciously, affirmed the sovereign authority of God and the healers. These therapies did not lead to rebels, revolutionaries, heretics, or ungovernable selves, except by accident. Here I am thinking of Martin Luther—an Augustinian monk who underwent spiritual direction (a commitment therapy)—who rebelled against the Catholic Church, though his and others' rebellions were not metaphysical. They still accepted the sovereignty of God, just not the pope's. Not surprisingly, the rebels asserted their authority, and commitment therapies remained unchanged at the core. Western commitment therapies, then, presupposed and unwittingly reinforced governable selves.

Ungovernable Selves and Psychoanalysis

There are, it seems to me, many possible positive motivations for listening to and helping people, such as to facilitate healing, greater autonomy/differentiation, mindfulness/consciousness, insight, freedom, and intimacy. There may also be underlying or subtle motivations, such as Fanon's helping patients become conscious of the real social sources of their suffering and choose an action in response to that suffering. Was Fanon hoping their actions would be rebellious or revolutionary toward the French colonizers? Was Fanon's therapeutic approach a modern form of commitment therapy? Perhaps, but I suspect he would have left choice or commitment to the patient because, if he had not, then psychoanalytic therapy would simply be yet another form of commitment therapy. If I am correct, Fanon and others like him were products of a type of education, informed by the rubrics of science, that did not serve the traditional apparatuses of religious sovereignty.[14] However, as I addressed in Chapter 1, features of sovereignty and the ontological rift were embedded in psychoanalytic theory and its attendant political philosophies. This said, psychoanalysis is not and was not aimed at forming good citizens, especially good citizens for a market society or even a patriarchal society. Bertha Pappenheim's trajectory after analysis is an illustration of this. In my view, psychoanalysis did not and does not have conscious allegiances to a state or religion. Psychoanalytic therapies, then, are not commitment therapies, which does not imply that patients' or therapists' commitments are not important and worth exploring. Indeed, I believe Fanon was concerned about his patients' commitments, especially their unconscious commitments to the very colonizing apparatuses that gave rise to their suffering. However, analytic therapies are not aimed at commitment to a particular religious or philosophical tradition.

In this section, I argue that psychoanalysis can be seen, in part, as a process that facilitates the development of ungovernable selves, by which I do not necessarily mean rebels or revolutionaries—people who have already demonstrated their passionate commitments and who may or may not be ungovernable selves. I also aim to indicate why ungovernable selves, whether facilitated by therapies or already present in other communities, are important when it comes to the climate crisis. Let me begin by returning to Chapter 2 and the brief discussion on Agamben's notion of impotentiality. Recall that Agamben (1999) illustrates potentiality in terms of impotentiality. He wrote:

> *Other living beings are capable only of their specific potentiality; they can only do this or that. But human beings are the animals who are capable of their own impotentiality. The greatness of human potentiality is measured by the abyss of human impotentiality.* Here it is possible to see how the root of freedom is to be found in the abyss of potentiality. To be free is not simply to have the power to do this or that thing, nor is it simply to have the power to refuse to do this or that thing. To be free is … *to be capable of one's own impotentiality.*
>
> (pp. 182–183, emphasis in original)

Agamben illustrates this by relying on Herman Melville's *Bartleby, the Scrivener*, wherein Bartleby was asked by his boss to do something, and Bartleby replied, "I prefer not to." For Agamben, Bartleby was exercising the freedom of not actualizing his potentiality, which means, in part, that Bartleby was not determined and could not be determined (in the sense of being commanded by others) to actualize his potentiality. Colebrook and Maxwell (2016) add that "to have potentiality is to be capable of not becoming what one has the capacity to be" (p. 38). In Chapter 2, I gave an example of a concert pianist who can prefer not to actualize her capacity to perform.

In this sense, Bartleby, in preferring not to, was not governable, even though he had to suffer the consequences of exercising his capacity for impotentiality. To dig a bit deeper, not every example of exercising impotentiality is necessarily a positive or healthy expression of ungovernability and freedom. Bartleby, in my view, was doing this mainly for himself, and his reasons for doing so were not entirely clear. His exercise of impotentiality lacked the aim of solidarity that Camus addressed. By contrast, in Chapter 1, I quoted Ruby Sales's poignant remarks about her parents. Ruby's parents were ungovernable in that they preferred not to operate out of the apparatuses of the oppressive, traumatizing sovereignty of white supremacy. They exercised their capacity for impotentiality aimed at their and their children's experiences of solidarity and dignity. Their capacity for impotentiality accompanied anarchic care. Ruby not only experienced this care but also recognized that she too could exercise her impotentiality in relation to the apparatuses of white supremacy. I suggest that Ruby and her parents were not rebels or revolutionaries but, as ungovernable selves, they, in solidarity, actively resisted the machinations and indignities of racism.[15]

I want to provide another example of ungovernable selves, but before I do let me return to psychoanalysis as a process of facilitating ungovernable selves. The notion of analytic love has been used frequently in psychoanalytic literature, along with its relation to analytic goals such as healing, emotional reparation, and insight. Daniel Shaw (2003; see also Davies, 2023; Hirsch, 1983), for instance, explores and summarizes the features of the notion of analytic love. Andrea Celenza (2007) also addresses the complexities of analytic love and its relation to power, responsiveness, and responsibility. While I find the notion worthwhile, I prefer the term "anarchic care" when referring to therapists' relationships with patients, not simply because it is more inclusive in that I genuinely care for my patients. It is also because I do not necessarily love all of them or believe that they need my love to heal or gain insight. In general, I prefer anarchic care because it entails an embodied, relational attunement and is aimed at exercising my analytic role for the resonant well-being of patients. Implicit here is the exercise of what I have previously called positive power in an asymmetrical relationship—a power that renders inoperative any apparatus of sovereignty or apparatuses of the ontological rift. This may seem obvious, but many past (and present) forms of commitment therapies were inextricably connected to apparatuses of sovereignty, as well as apparatuses that produce the ontological rift. In contrast, as in Fanon's care for his patients in Algeria or a German psychoanalyst treating patients during the Nazi era, anarchic care would entail rendering the apparatuses of tyrannical sovereignty inoperative, whether in the form of French colonization or Nazi oppression (or current forms of authoritarianism). Anarchic care renders these apparatuses inoperative, creating a relational space for the possibility of patients exercising their impotentiality—of becoming ungovernable selves. As I mentioned above, I am confident that

Fanon was not interested in using therapy to encourage people to be rebels or revolutionaries. Rather, through his anarchic care, he invited them to see the real social sources of their suffering and to choose an action toward those apparatuses. While Fanon was interested in patients choosing an action, it was up to patients what action and commitment to choose. To choose an action is already a sign of a commitment.

Another example of ungovernable selves provides a different angle before this is connected to the climate crisis. Throughout this book, I have endeavored to think with Agamben and Indigenous peoples. In the previous section, I mentioned the plight of Indigenous peoples in the United States, indicating that, despite innumerable attempts to subjugate and assimilate them into Western notions of nationalism and sovereignty, they resisted in a variety of ways, including rebellion. They consistently proved they were ungovernable. In general, they organized and continue to organize their societies without using apparatuses of sovereignty and sovereign classes. Moreover, Indigenous apparatuses do not produce the ontological rift and dissonant subjects. Rather, the stories, rituals, and practices of Indigenous peoples are concerned with establishing and maintaining resonant relationships with each other, other species, and the land. Resonant relations imply anarchic care, responsibility, and commitment to the exercise of positive power, which includes rituals or practices of repair of dissonant relations. In my view, the very exercise of Indigenous apparatuses represents the reality of ungovernable selves and attendant forms of resistance to the dissonant relations produced by Western apparatuses of the rift that aim to control, manage, assimilate, or subjugate othered human beings, other species, and the Earth. Put differently, in Indigenous anthropologies dissonance is an epiphenomenon of resonance, while in Western anthropologies it is the reverse. One hypothesis is that the antipathy of white Euro-Americans toward Indigenous peoples was connected to how Indigenous people represented a way of life that Westerners had abandoned for the sake of control over nature. Rather than become aware of and take responsibility for the unconscious emptiness and futility that result from being alienated from what Westerners called "nature," colonizers reviled, controlled, subjugated, and often eliminated Indigenous peoples. Meanwhile, Indigenous peoples proved time and again that they and their apparatuses for organizing social relations were and are not governable. Indeed, Indigenous ways of life were (and are) inimical, as Frank Linderman (Lear, 2006) came to realize, to Western ways of being in the world.

This detour enables me to note further the connection between ungovernable selves, resonance, and analytic therapies. I do not think an aim of therapy is merely to facilitate patients' capacity for exercising their impotentiality in the sense of Bartleby. As suggested above, what Agamben leaves out in his analysis of impotentiality are the ends of "preferring not to." One end could simply be a liberal or individualistic view of freedom, as in the case of

Bartleby. Another might be Camus's contention that solidarity and dignity are the aims of rebellion and revolution—preferring not. Yet another might be freedom from being in bondage to traumatic memories. I remember numerous patients getting to a point in therapy when they said, in various ways, they preferred not to succumb to their past traumas. The process of therapy helped create a space where patients had more psychological differentiation and freedom to exercise inoperativity with regard to past traumas. Another aim might be a patient who prefers not to remain in an abusive relationship, exercising their freedom to live a life of self-respect, self-confidence, and self-esteem. These are admirable results of an analytic process and patients' work. Yet, I wish to add another way to think about the aim of psychoanalytic therapies' facilitation of ungovernable selves, and that is through the concept of resonance.

The aim of facilitating ungovernable selves is resonant relations of belongingness wherein persons experience forms of anarchic care such that mutual interactions between selves who feel respected and worthwhile exist. I mentioned above Herman's (1992) view that traumatic events "destroy the sustaining bonds between the individual and community" (p. 214), whereas therapy aims to facilitate patients' finding sustaining communities of resonant relations of solidarity. Another way to say this is that therapy facilitates patients' engagement in resonant relations not only with the therapist but also with members of the self-chosen communities they are committed to and where they experience solidarity with others who recognize and respect their singularities.

But I want to try to close the loop on psychoanalysis and Searles's encouragement to include other species in our views of psychosocial development and practice. I believe one of the things therapists need to consider about themselves and their patients is their resonant (or not) relationships with other species and the Earth. If the climate polycrisis has revealed the fundamental architectural problem of Western ways of being in the world, then it seems we need to consider, as this book attests, how this impacts theory and practice. In my view, one aim of psychoanalysis is the facilitation of persons' exercise of their capacity for impotentiality, which has as its further aim resonant solidarity, not simply with other human beings in a sustaining community but also with other species. Naturally, this implies that communities of which we are a part also have as their aims and practices resonant relations with other species. In other words, these are ungovernable selves and communities that prefer not to live out of the apparatuses of the rift and instead create spaces to live resonantly with other species.

This may appear more in line with commitment therapies, but that is not the case. Like Fanon's approach, this view entails facilitating the process of recognizing the social, political, and economic sources of suffering and then choosing actions toward those sources. Patients choose which communities, organizations, and so on, to commit to and how to live resonantly with other human beings, other species, and the Earth. Naturally, all of this depends on patients' needs, capacities, personal aims, and, if they are dealing with trauma, where they are in the process of working through it. We may never get to a place of exploring their beliefs, attitudes, and behaviors toward other species and the Earth. However, for those patients who are keen to explore these topics, this is a possible road forward, especially for patients who are more aware of their eco-emotions regarding the climate crisis.

All of this said, I believe strongly that therapists, analytic or otherwise, deal with patients' vulnerabilities, and this means confronting our own vulnerabilities and ways of being in the world. What follows from this is the need to consider (1) our resonant or dissonant relations with other species and the Earth and (2) our commitments to communities and organizations to the extent to which they facilitate resonant relations. In terms of the rift, to what extent are we or our communities and organizations ungovernable, preferring not to live out of the apparatuses of the rift, while creating spaces to live more resonantly with what we call "nature"?

I wish to stress that this view does not mean psychoanalytic therapies foster rebels and revolutionaries. To be sure, as noted above, there seems to be some evidence of rebels appearing frequently in psychoanalytic history. But I have also pointed out that a rebel or revolutionary may be a governable self. Whenever we read in psychoanalytic history of an analyst wedded to their revolutionary theory (and operating out of the rift), then we can be confident there is a governable self (in my view, Freud, Lacan, and Klein are examples). This is not to suggest that these men and women have not made contributions but rather that, as Phillips suggests, the theory becomes an idol, and all idols have committed or devoted governable selves. My view of analytic therapies and their aims, then, is that there are ways of reimaging psychoanalysis given the climate crisis. It is a way of thinking that is not the Truth, but may, I hope, have seeds of truth that can help psychoanalysts work with patients who recognize and experience distress in relation to the climate crisis and who, in this process, exercise their capacities for impotentiality aimed at establishing and committing to resonant relations with other human beings, other species, and the Earth.

Souls, Spirits, and Psychoanalysis

I would like to acknowledge psychoanalysis's history of a difficult, ambivalent, and often antagonistic relationship with religion because I believe there is a path toward some shared anthropological aims. It is important to situate this ambivalent relationship within the rise of modernity in the 19th century. In so doing, divergences become evident in analytic versus commitment therapies. I believe there is a way to bridge this divide, especially if we think with Indigenous philosophies that can, as addressed in Chapter 2, alter how we understand psychosocial development and our relations with the nonhuman world.

Jacques Derrida (2008) recognized the "abyssal rupture" (p. 30) between human beings and other species. This rupture, which he located in Western philosophies and the Abrahamic traditions, involves constructing other species in terms of lacking this or that feature of the human. Derrida wrote:

> It is *not just* a matter of asking whether one has the right to refuse the animal such and such a power (speech, reason, experience of death, mourning, culture, institutions, technics, clothing, lying, pretense of pretense, covering of tracks, gift, laughter, crying, respect, etc.—the list is necessarily without limit and the most powerful philosophical tradition in which we have lived has refused the "animal" *all of that*). It also means asking whether what calls itself human has the right rigorously to attribute to man, which means therefore to attribute to himself, what he refuses the animal, and whether he can ever possess the *pure, rigorous, indivisible* concept, as such, of that attribution.
>
> (p. 135, emphasis in original)

We can include in this list of "deficiencies" the notion of spirit or soul. Other species have no souls, the Abrahamic traditions believe. Other species, these religious traditions believe, reside neither in heaven nor hell after death. We can conclude from this that they are deemed to be without ontological significance while the ontological significance of human beings—usually only *some* human beings—is affirmed.

One might point out that Aristotle (and other philosophers) believed animals have souls, but this was based on a taxonomic hierarchical scale that placed human beings on top. Other species have souls, but they nevertheless, for Aristotle, lack intelligence, agency, reason, and so on. The question, in my view, is not whether other species have souls or spirits. Rather, the question is what the appellation "soul" refers to and how it is used. In other words, these terms (soul, spirit) are abstractions and are not real in the sense that they are not quantifiable realities. Nevertheless, they are real to the extent that using these abstractions is rooted in human experiences and has psychosocial functions and consequences. For instance, because we deny other species have souls (or believe they have lesser, nonnormative souls), Western subjects have either instrumental experiences in relation to other species or no conscious experiences at all, only the normative, political, and historical unconscious. There is, then, no experience of resonant togetherness, only separation and alienation. The terrible ecological consequences of this I have already pointed out in previous chapters. From a different angle, other species, who are perceived as soulless, lack singularity and existential significance since possessing a soul implies singularity and significance. As noted in previous chapters, this functions to assure Western subjects of their ontological significance while at the same time justifying and legitimating their instrumental use and exploitation of other species. Moreover, as Byung-Chul Han (2022) notes,

> We exploit the world with such brutality because we have declared matter to be something dead and the earth to be a mere resource. … What is needed is an altogether different understanding of the earth and matter. … Thus, ecology must be preceded by a new ontology of matter, one that views it as something that lives.
>
> (p. 96)

By contrast, most Indigenous peoples' ecology affirms not only that other species have souls but that the land itself is connected to a great or vast spiritual reality. In believing other species have souls, Indigenous peoples acknowledge and respect the multiple intelligences and singularities of other species. This is what the signifier or abstraction "spirit" means in these contexts. This abstraction is also connected to how Indigenous peoples experience other species and the Earth, as well as how they behave toward them. Indigenous peoples, generally speaking, experience a sense of resonant belonging or togetherness with regard to other species and the Earth, which is reflected in their behaviors toward both. To say and to believe that other species have souls and that the Earth is sacred entail dispositions and behaviors committed to respecting, caring for, and acting justly toward all who are kin (e.g., Deloria, 2003; Lloyd, 2024). Yes, Indigenous peoples kill other species and can have antagonistic relations with some species, but this does not mean these other species are perceived to be without souls (nor, in the Aristotelean sense, to have lesser souls).

To consider these beliefs to be beneficial or even ecological illusions is not my point. Indeed, I think it is a mistake to categorize them as such because the notion of "illusion" means a mistaken belief. Also, to do so follows the tradition of Western colonization with its arrogance and self-certainty. I prefer instead to focus on the beliefs or abstractions people possess and how these are related to their experiences of, relations to, and behaviors toward other human beings, other species, and the Earth. What are the consequences of these collective and individual beliefs? Do these religious (or philosophical) beliefs contribute to resonant relations and experiences of belonging or togetherness, or do they reinforce dissonant relations characterized by alienation, dissociation, and violence? From my perspective, these are therapeutic considerations. That is, psychoanalytic therapists, while interested in bringing the unconscious to consciousness, are also concerned with patients' relationships and how they contribute to or undermine their flourishing. We are, after all, social and communal animals. And yet, as the climate crisis has shown, we are also animals who are materially and existentially dependent on the biodiverse Earth. This means that therapists should, in some cases, be interested in patients' experiences and their relations to other species and the Earth. This entails the patients' conscious and unconscious beliefs, spiritual or otherwise, about other species and the Earth. Of course, I think therapists have a public role as well and are invited to delve into social, political, and economic discourses that have implications for the environment. And, since psychoanalytic therapies are premised on therapists undergoing their own analyses, then we need to consider critically our beliefs, abstractions, and relations regarding other species and the Earth.

Does this suggest that therapists or scientists or Western religious persons need to adopt Indigenous beliefs that other species have souls and the land has a spirit? I am in agreement with Eduardo Viveiros de Castro (2017, p. 196), who believed that, though we (Westerners) cannot think like Indigenous people, we can think with them. If a scientist or therapist cannot accept the idea that other species have souls, fair enough. To try to believe something that does not fit with Western sciences can be problematic. Nevertheless, Westerners can believe that other species possess singularities and intelligences. If no two snowflakes are identical (each is singular), then no two animals from the same species are identical. If we can accept the singularity of other human beings, it is not unscientific or romantic to accept the reality of the singularities and multiple intelligences of other species. This is not merely a convenient abstraction. Rather, it is a genuine belief that can shape how we see and relate to other species and the Earth. Perhaps we will experience a shared sense of belonging or togetherness with these manifold creatures of multiple intelligences that share with us this singular habitat we call Earth. Maybe we will no longer wonder if we are alone and lonely in this vast universe but will recognize that we share the world with innumerable creatures who have varied and remarkable intelligences. It might even be possible to be curious about, listen to, and learn from other species. And maybe all of this will be a step toward an ecological, therapeutic approach that aims toward ungovernable selves who experience resonant belongingness with our kin. And these resonant experiences of ungovernable selves can accompany our commitments to recognize and respect the singularities of other living beings.

Conclusion

Theories about psychosocial development and therapy are modern-day abstract attempts to tell stories. Many of us modern (or postmodern) people grew up believing that human science theories tend to be closer to the truth or fact than novels or myths. We have, I believe, more freedom and intellectual imagination when we believe that our theories are simply stories we tell rather than the Truth or fact. To be sure, novels and other stories can reveal truths about being human (or otherwise), but we tend to focus more on the aesthetics of a story rather than on whether it is the Truth. Indigenous peoples tell numerous stories without devolving into intellectual straitjackets of doctrine and dogma that become attached to cherished theories. But I wonder if we modern Westerners have a secularized version of religious truth lingering in our view of scientific theories or our understanding of scientific theories. Political legal philosopher (and unrepentant Nazi) Carl Schmitt noted, "All significant concepts of the modern theory of the state are secularized theological concepts" (as quoted in Brown, 2015, p. 59). I suspect this is also true, at times, for how we regard our prized theories. We become attached to our theories, and this is often reinforced by those who believe and make use of the same theories. To be convinced of this, read the history of psychoanalysis (or scientific theories) and reflect on the "religious" wars waged against non-believers. Freud, Jung, Klein, Lacan, and others acted as if their theories were true, and, if not the Truth, something that approximated it.

I am not suggesting we rid ourselves of notions of fact and truth. I am, however, implying that the previous chapters are my way of telling a story. Is my story better, more factual, and truer than other analytic theories? That is not my question. In thinking with Agamben and Indigenous peoples, I am curious about whether a story or theory helps situate us ecologically in relation to other human beings, other species, and the Earth. As Indigenous peoples have shown, there is no single story. Indeed, given the multiplicity of singularities, intelligences, and habitats, there are numerous stories to tell. This book is one story or theory that seeks to incorporate an ecological foundation of psychoanalytic theory and concepts. Readers, with their multiple critical intelligences, will decide whether this theory approximates a resonant ecological way of understanding human beings and, more particularly, psychosocial development and therapy. Even better, readers can come up with their own stories or theories aimed at overcoming the rift and becoming ungovernable ecological selves committed to anarchic, resonant care of human beings, other species, and the Earth. Let Erich Fromm (1973) have the last word about what is, in my view, the aim of anarchic care and the resonant belonging of ungovernable selves: "The passionate love of life and of all that is alive … is the wish to further growth, whether in a person, a plant, an idea, or a social group" (p. 366).

Notes

1 S. Walsh, "Gov. Ron DeSantis signs law erasing climate change from Florida policy," UPI, May 16, 2024. Available at: www.upi.com/Top_News/US/2024/05/16/florida-governor-Ron-DeSantis-erase-climate-change-legislation/8891715830982/ (accessed December 4, 2025).

2 I. James & H. Smith, "'It's a huge loss': Trump administration dismisses scientists preparing climate report," *Los Angeles Times*, April 29, 2025. Available at: www.msn.com/en-us/news/politics/it-s-a-huge-loss-trump-administration-dismisses-scientists-preparing-climate-report/ar-AA1DSl7e?ocid=BingNewsSerp (accessed December 4, 2025).

3 The Nazis were deliberate in undermining the ethical foundations of therapy for the sake of creating governable selves. While democratic nations do not do this, they can reproduce governable selves. For instance, Western notions of a bounded, atomized self fit well with the needs of capitalist-driven society and culture (Zaretsky, 1976). Also, as Phillip Cushman (1995) argued, "Psychotherapy is permeated by the philosophy of self-contained individualism, exists within the framework of consumerism, speaks the language of self-liberation, and thereby unknowingly reproduces some of

the ills it is responsible for healing" (p. 6). Furthermore, Cushman noted, psychotherapy perpetuates the political status quo (cf. Zaretsky, 2005, p. 164).

4 Science can, at times, similarly function as an idol in the sense that it "presumes completeness of knowledge and by extension disciplinary completeness" (Gordon, 2023, p. 156). This is especially the case in Cartesian–Baconian sciences that seek to gain mastery over nature, which has included scientific assertions masquerading as facts, such as women's inferiority in relation to men, black persons' inferiority in relation to white people, and the inferiority of other species in relation to (Western) human beings.

5 Ellenberger (1970) and Frank (1961) identify a wide variety of rituals used throughout the world by shamans and other religious healers to respond to different types of human suffering. While there is variation within Western Christianity, practitioners did use a common text, the Bible, to support and justify their interpretations and interventions.

6 This does not include various rituals of exorcism, which involved casting out demons. Answering methods or talking cures were more common than exorcisms (Holifield, 1983).

7 An interesting example of classical therapies situated in a community is the town of Geel in the Netherlands. Since the 13th century, long before modern psychiatry, this town and its families accepted psychologically distressed persons. This is a practice that has continued today, though with significant reliance on the psychiatric sciences.

8 Granted, there are religious rituals for initiation, but these are not the ones identified in the cure of souls tradition for caring for the sick. Instead, healers and sufferers used the religious interpretive framework to provide meaning and to help restore or maintain an individual's relationship with the community and with God.

9 Recall that Hartmut Rosa (2019) claims that "the primal form of existence is a relationship not of alienation but of resonance" (p. 258).

10 One does not have to agree with various theological premises or stories to see that beneath the stories is an anthropology that points to the necessity of committed relationships for the survival and flourishing of individuals.

11 Christian history contains illustrations of horrific practices that rationalized political and religious violence with the ostensible aim of preserving both the community and an individual's soul (see Foucault, 1979; Giroux, 2012a). This said, I am addressing the ideal aims of classical therapies, which includes their shortcomings.

12 Lest we too narrowly judge our predecessors, they were limited in their ability to understand the physical etiologies of mental illnesses, which translated into limited ways of responding. Let me add that we continue to sequester people who suffer from mental illness, whether that is more formally in prisons and mental hospitals or in less formal settings such as inner-city streets (Foucault, 1979; Szasz, 1977).

13 It is interesting to note that metaphysical rebellion, for Camus (1956/1991), leads to the realization that we are alone (p. 10), which I assume means that human beings are alone in the cosmos. I suggest that this could only have come from a modern subject of the ontological rift. Modernity undermined many persons' confidence in an all-powerful God or God at all. This, combined with the rift's belief that other species lack singularity and intelligence (and are not part of the polis except in instrumental ways), means that without God we are indeed alone. I do not believe Indigenous philosophers, with their beliefs in a multiplicity of singularities and intelligences, would arrive at that conclusion. They believe they are surrounded by kin, including the land.

14 A related question regards the degree to which modern therapies unwittingly serve as apparatuses of sovereignty, reinforcing governable selves.

15 There are several points here. First, persons who self-identify as rebels and revolutionaries are not necessarily ungovernable selves, as noted above. Second, self-identified rebels and revolutionaries can be ungovernable selves as long as they do not replace one form of sovereignty with another. Third, the state apparatuses can label people as rebels or revolutionaries for purposes of control, subjugation, or annihilation. Finally, Ruby later self-identified as an activist, rather than a rebel or revolutionary. Her activism, while certainly a kind of rebellion or revolution against white supremacy, is, in my view, better understood as an expression of her ungovernable self.

References

Adler, E., & Bachant, J. L. (1998). Intrapsychic and interactive dimensions of resistance: A contemporary perspective. *Psychoanalytic Psychology*, 15, 451–479. doi:10.1037/0736-9735.15.4.451.

Agamben, G. (1998). *Homo sacer: Sovereign power and bare life* (D. Heller-Roazen, Trans.). Stanford University Press.

Agamben, G. (1999). *Potentialities: Collected essays in philosophy* (D. Heller-Roazen, Trans.). Stanford University Press.

Agamben, G. (2004). *The open: Man and animal* (K. Attell, Trans.). Stanford University Press.

Agamben, G. (2005). *State of exception*. Stanford University Press.

Alexander, M. (2010). *The new Jim Crow: Mass incarceration in the age of colorblindness*. The New Press.

Alford, C. F. (2013). *Trauma and forgiveness: Consequences and communities*. Cambridge University Press.

Altman, N. (2000). Black and white thinking: A psychoanalyst reconsiders race. *Psychoanalytic Dialogues*, 10, 589–605. doi:10.1080/10481881009348569.

Altman, N. (2004). Whiteness uncovered: Commentary on papers by Melanie Suchet and Gillian Straker. *Psychoanalytic Dialogues*, 14, 439–446. doi:10.1080/10481881409348797.

Amery, J. (1995). Torture. In L. Langer (Ed.), *Art from the ashes: A Holocaust anthology* (pp. 121–136). Oxford University Press.

Anderson, C. (2016). *White rage: The unspoken truth of our racial divide*. Bloomsbury.

Applegate, J. S. (1989). The transitional object reconsidered: Some sociocultural variations and their implications. *Child and Adolescent Social Work*, 6(1), 38–51. doi:10.1007/BF00755709.

Aralepo, O. (2003). The white male therapist/helper as (m)other to the black male patient/client: Some intersubjective considerations. *Free Associations*, 10, 382–398.

Arendt, H. (1958). *The human condition*. University of Chicago Press.

Aristotle. (1971). *The politics of Aristotle* (E. Barker, Ed. & Trans.). Oxford University Press.

Aristotle. (1987). *De anima* [On the soul] (H. Lawson-Tancred, Trans.). Penguin.

Aron, L., & Mitchell, S. A. (1999). *Relational psychoanalysis: The emergence of a tradition*. Analytic Press.

Aspey, L., Jackson, C., & Parker, D. (2023). *Holding the hope: Reviving psychological and spiritual agency in the face of climate change*. PCCS Books.

Augustine. (1963). *The confessions of St. Augustine* (R. Warner, Trans.). Mentor Books.

Bacal, H. A. (Ed.). (1998). *Optimal responsiveness: How therapists heal their patients.* Aronson.

Baptist, E. E. (2014). *The half has never been told: Slavery and the making of American capitalism.* Basic Books.

Barad, K. (2007). *Meeting the universe halfway: Quantum psychics and the entanglement of matter and meaning.* Duke University Press.

Battiste, M. (2013). *Decolonizing education: Nourishing the learning spirit.* UBC Press.

Bauer, G. P. (1993). *The analysis of the transference in the here and now.* Aronson.

Bauer, G. P. (Ed.). (1994). *Essential papers on transference analysis.* Aronson.

Becker, E. (1997). *The denial of death.* Free Press.

Bednarek, S. (Ed.). (2024). *Climate, psychology and change: Reimagining psychotherapy in an era of global disruption and climate anxiety.* North Atlantic Books.

Beebe, B., & Lachmann, F. M. (1994). Representation and internalization in infancy: Three principles of salience. *Psychoanalytic Psychology,* 11(1), 127–165. doi:10.1037/h0079530.

Beebe, B., & Lachmann, F. M. (2002). *Infant research and adult treatment: Co-constructing interactions.* Analytic Press.

Benjamin, J. (1995). Sameness and difference: Toward an "overinclusive" model of gender development. *Psychoanalytic Inquiry,* 15, 125–142. doi:10.1080/07351699509534021.

Benjamin, J. (1996). In defense of gender ambiguity. *Gender and Psychoanalysis,* 1, 27–43.

Birch, J. (2024). *The edge of sentience: Risk and precaution in humans, other animals, and AI.* Oxford University Press.

Black, D. (2015). *The coming.* St. Martin's Press.

Blanchot, M. (1995). *The writing of the disaster* (A. Smock, Trans.). University of Nebraska Press.

Bodin, J. (2009). *On sovereignty: Six books of the commonwealth.* Seven Treasures.

Bodnar, S. (2024a). The fierce urgency of now: The case for including the environmental field in clinical work. *Psychoanalytic Dialogues,* 34, 795–811. doi:10.1080/10481885.2024.2416621.

Bodnar, S. (2024b). Sensitivity and birds in a red sky: Finding relationality in an environmental dialogue—reply to Federici, Somers, and Stothart. *Psychoanalytic Dialogues,* 34, 833–838. doi:10.1080/10481885.2024.2417606.

Bollas, C. (1987). *The shadow of the object: Psychoanalysis of the unthought known.* Columbia University Press.

Boluda-Verdú, I., Senent-Valero, M., Casas-Escolano, M., Matijasevich, A., & Pastor-Valero, M. (2022). Fear for the future: Eco-anxiety and health implications, a systematic review. *Journal of Environmental Psychology,* 84, 1–17. doi:10.1016/j.jenvp.2022.101904.

Bonovitz, C., & Harlem, A. (Eds.). (2018). *Developmental perspectives in child psychoanalysis and psychotherapy.* Routledge.

Bookchin, M. (2005). *The ecology of freedom: The emergence and dissolution of hierarchy.* AK Press.

Bookchin, M. (2023). *Remaking society: A new ecological politics.* AK Press.

Boothby, R. (2023). *Embracing the void: Rethinking the origin of the sacred.* Northwestern University Press.

Breger, L. (2000). *Freud: Darkness in the midst of vision.* John Wiley.
Bremer, J. D., & Marmar, C. R. (1998). *Trauma, memory, and dissociation.* American Psychiatric Association Press.
Brickman, C. (2018). *Aboriginal populations in the mind: Race and primitivity in psychoanalysis.* Routledge.
Brody, S. (1980). Transitional objects: Idealization of a phenomenon. *Psychoanalytic Quarterly*, 49, 561–605.
Bromberg, P. (2001). Treating patients with symptoms—and symptoms with patience: Reflections on shame, dissociation, and eating disorders. *Psychoanalytic Dialogues*, 11(6), 891–912. doi:10.1080/10481881109348650.
Brown, W. (2010). *Walled states, waning sovereignty.* Zone Books.
Brown, W. (2015). *Undoing the demos: Neoliberalism's stealth revolution.* Zone Books.
Brown, W. (2023). *Nihilistic times: Thinking with Max Weber.* Belknap.
Bubeck, D. (1995). *Care, gender, and justice.* Clarendon Press.
Buber, M. (1958). *I and thou.* Charles Scribner.
Bunge, M. J. (2001). *The child in Christian thought.* Erdmanns.
Cambray, J. (2017). The emergence of the ecological mind in Hua-Yen/Kegon Buddhism and Jungian Psychology. *Journal of Analytical Psychology*, 62, 20–31.
Camus, A. (1991). *The rebel.* Vintage. (Original work published 1956)
Canessa, A., & Picq, M. L. (2024). *Savages and citizens: How indigeneity shapes the state.* University of Arizona Press.
Carrette, J., & King, R. (2005). *Selling spirituality: The silent takeover of religion.* Routledge.
Carson, R. (2002). *Silent spring.* Houghton Mifflin. (Original work published 1962)
Caruth, C. (1996). *Unclaimed experience: Trauma, narrative, and history.* Johns Hopkins University Press.
Caruth, C. (2014). *Listening to trauma: Conversations with leaders in the theory and treatment of catastrophic experience.* Johns Hopkins University Press.
Castoriadis, C. (1997). *World in fragments: Writings on politics, society, psychoanalysis, and the imagination.* Stanford University Press.
Celenza, A. (2007). Analytic love and power: Responsiveness and responsibility. *Psychoanalytic Inquiry*, 27, 287–301. doi:10.1080/07351690701389478.
Chessick, R. D. (1991). *The technique and practice of intensive psychotherapy.* Aronson.
Chibber, V. (2013). *Postcolonial theory and the specter of capital.* Verso.
Cholbi, M. (2021). *Grief: A philosophical guide.* Princeton University Press.
Chopra, D., Penrose, R., Carter, B., Stapp, H. P., Hameroff, S., Dobyns, Y. H., Kafatos, M., Mensky, M. B., Globus, G., & Kak, S. (2017). *How consciousness became the universe: Quantum physics, cosmology, neuroscience, parallel universes.* Science.
Clebsch, W. A., & Jaekle, C. R. (1994). *Pastoral care in historical perspective.* Aronson.
Clough, P. T. (2024). Bring in the social, bring in the noise. *Psychoanalytic Dialogues*, 34, 522–532.
Coates, T.-N. (2015). *Between the world and me.* Spiegel & Grau.
Coburn, W. J. (2001). Subjectivity, emotional resonance, and the sense of the real. *Psychoanalytic Psychology*, 18, 303–319. doi:10.1037/0736-9735.18.2.303.
Coburn, W. J. (2024). Freedom and self-ownership: An emergence theory of free will. *Psychoanalysis, Self, and Context*, 19, 139–156. doi:10.1080/24720038.2023.2281404.

Cocks, G. (2018). *Psychotherapy in the Third Reich* (2nd ed.). Routledge. (Original work published 1985)

Colebrook, C., & Maxwell, J. (2016). *Agamben*. Polity.

Coles, R. (1975). *The mind's fate: Ways of seeing psychiatry and psychoanalysis.* Little, Brown.

Colman, A. V. (2022). Corporeal schemas and body image: Fanon, Merleau-Ponty, and the lived experience of race. In L. Laubscher, D. Hook, & M. U. Desai (Eds.), *Fanon, phenomenology, and psychology* (pp. 127–138). Routledge.

Comtesse, H., Ertl, V., Hengst, S. M. C., Rosner, R., & Smid, G. E. (2021). Ecological grief as a response to environmental change: A mental health risk or functional response? *International Journal of Environmental Research and Public Health*, 18, Article 734. doi:10.3390/ijerph18020734.

Cone, J. H. (1970). *A black theology of liberation.* Orbis Books. (Original work published 1970)

Cox, H. (2016). *The market as God.* Harvard University Press.

Cozolino, L. (2006). *The neuroscience of human relationships: Attachment and the developing social brain.* Norton.

Crockett, C. (2012). *Radical political theology: Religion and politics after liberalism.* Columbia University Press.

Crutzen, P. J., Stoermer, E. F., & Steffen, W. (2000). The "Anthropocene." *IGB Global Change Newsletter*, 41, 17–18. Available at: www.jstor.com/stable/j.ctt5vm5bn.52.

Cunsolo, A., Borish, D., Harper, S. L., Snook, J., Shiwak, I., & Wood, M. (2020). "You can never replace the caribou": Inuit experiences of ecological grief from caribou declines. *American Imago*, 77, 31–59. doi:10.1353/aim.2020.0002.

Cunsolo, A., Harper, S. L., Minor, K., Hayes, K., Williams, K. G., & Howard, C. (2020). Ecological grief and anxiety: The start of a healthy response to climate change? *The Lancet Planetary Health*, 4, 261–263. doi:10.1016/S2542-5196(20)30144-30143.

Cunsolo, A., & Landman, K. (2017). *Mourning nature: Hope at the heart of ecological loss and grief.* McGill-Queen's University Press.

Cushman, P. (1995). *Constructing the self, constructing America: A cultural history of psychotherapy.* Addison Wesley.

Dalal, F. (2002). *Race, colour and the process of racialization: New perspectives from group analysis, psychoanalysis, and sociology.* Brunner-Routledge.

Danner, M. (2009). *Stripping bare the body: Politics, violence, war.* Nation Books.

Davies, J. (2023). Reawakening desire: Shame, mourning, analytic love, and psychoanalytic imagination. *Psychoanalytic Dialogues*, 33, 285–301. doi:10.1080/10481885.2023.2204773.

De Vos, R. (Ed.). (2023). *Decolonising animals.* Sydney University Press.

DeCaroli, S. (2007). Boundary stones: Giorgio Agamben and the field of sovereignty. In M. Calarco & S. DeCaroli (Eds.), *Giorgio Agamben: Sovereignty and life* (pp. 43–69). Stanford University Press.

DeCasper, A., & Fifer, W. (1980). Of human bonding: Newborns prefer their mothers' voices. *Science*, 208, 1174–1176.

DeCasper, A., & Spence, M. (1986). Prenatal maternal speech influences newborns' perception of speech sounds. *Infant Behavior and Development*, 4, 19–36. doi:10.1126/science.7375928.

Deloria, V., Jr. (1997). *Red earth, white lies: Native Americans and the myth of scientific fact.* Fulcrum.

Deloria, V., Jr. (2003). *God is red: A native view of religion*. Fulcrum.
Derrida, J. (2008). *The animal that therefore I am*. Fordham University Press.
Des Pres, T. (1976). *The survivor: An anatomy of life in the death camps*. Oxford University Press.
Desmond, M. (2016). *Evicted: Poverty and profit in the American city*. Crown.
Desmond, M. (2023). *Poverty, by America*. Crown.
deVries, M. (1996). Trauma in cultural perspective. In B. A. van der Kolk, A. C. McFarlane, & L. Weisaeth (Eds.), *Traumatic stress: The effects of overwhelming experience on mind, body, and society* (pp. 398–416). Guilford.
Dickens, C. (2013). *A Christmas carol*. Animedia.
Dickinson, C. (2015). The "absence" of gender. In C. Dickinson & A. Kotsko (Eds.), *Agamben's coming philosophy: Finding a new use for theology* (pp. 167–182). Rowman & Littlefield.
Dodds, J. (2012). *Psychoanalysis and ecology at the edge of chaos: Complexity theory, Deleuze/Guattari and psychoanalysis for a climate in crisis*. Routledge.
Dodds, J. (2019). Otto Fenichel and ecopsychoanalysis in the Anthropocene. *Psychoanalytic Perspectives*, 16(2), 195–207. doi:10.1080/1551806X.2019.1601921.
Dostoevsky, F. (2009). *The brothers Karamazov*. Lowell Press.
Du Bois, W. E. B. (2003). *The souls of black folk*. Barnes & Noble Books. (Original work published 1903)
Dufour, D.-R. (2008). *The art of shrinking heads: On the new servitude of the liberated in the age of total capitalism* (D. Macey, Trans.). Polity Press.
Dussel, E. (1985). *Philosophy of liberation*. Orbis Books.
Eagleton, T. (2016). *Materialism*. Yale University Press.
Eberhart, R. (2024). Christian animist economics. In S. Oppermann & S. Iovino (Eds.), *Environmental humanities: Voices from the Anthropocene* (pp. 137–153). Rowman & Littlefield.
Ellenberger, H. F. (1970). *The discovery of the unconscious: The history and evolution of dynamic psychiatry*. Basic Books.
Engster, D. (2007). *The heart of justice: Care ethics and political theory*. Oxford University Press.
Engster, D., & Hamington, M. (2015). *Care ethics and political theory*. Oxford University Press.
Engstrom, S. (2019). Recognizing the role eco-grief plays in responding to environmental degradation. *Journal of Transdisciplinary Peace Praxis*, 1(1), 168–186.
Epstein, H. (1988). *Children of the Holocaust: Conversations with sons and daughters of survivors*. Penguin.
Epstein, L. J., & Feiner, A. (1993). *Countertransference: The therapists' contribution to the therapeutic situation*. Aronson.
Erdoes, R., & Ortiz, A. (Eds.). (1984). *American Indian myths and legends*. Pantheon.
Erikson, E. H. (1982). *The life cycle completed*. W. W. Norton.
Faber, R. (2023). *The mind of Whitehead: Adventure in ideas*. Pickwick.
Fanon, F. (2008). *Black skin, white masks*. Grove Press. (Original work published 1952)
Fassin, D., & Rechtman, R. (2009). *The empire of trauma: An inquiry into the condition of victimhood*. Princeton University Press.
Feuerbach, L. (1989). *The essence of Christianity* (G. Elliot, Trans.). Prometheus Books.
Fingarette, H. (2000). *Self-deception*. University of California Press. (Original work published 1969)

Flew, A. (1978). Transitional objects and phenomena: Interpretations and comments. In S. A. Grolnick & L. Barkin (Eds.), *Between reality and fantasy: Transitional objects and phenomena* (pp. 483–502). Aronson.

Fonagy, P. (2001). *Attachment theory and psychoanalysis.* Other Press.

Fonagy, P., & Target, M. (1997). Attachment and reflective function: Their role in self–organization. *Development and Psychopathology*, 9, 679–700. doi:10.1017/S0954579497001399.

Foucault, M. (1965). *Madness and civilization: A history of insanity in the age of reason.* Vintage Books.

Foucault, M. (1972). *Power/knowledge: Selected interviews and other writings, 1972–1977.* Pantheon Books.

Foucault, M. (1979). *Discipline and punishment: The birth of the prison.* Vintage Books.

Foucault, M. (1987). *Mental illness and psychology.* University of California Press. (Original work published 1954)

Fraisart, S. C. (2021). *The liberalism of care: Community, philosophy and ethics.* University of Chicago Press.

Frank, G. (2012). On the concept of resistance: Analysis and reformulation. *Psychoanalytic Review*, 99, 421–435.

Frank, J. D. (1961). *Persuasion and healing: A comparative study of psychotherapy.* Schocken Books.

Frank, T. (2000). *One market under God: Extreme capitalism, market populism, and the end of economic democracy.* Anchor Books.

Frankfurt, H. G. (2005). *On bullshit.* Princeton University Press.

Fraser, N. (2020). *Fortunes of feminism: From state-managed capitalism to neoliberal crisis.* Verso.

Fraser, N. (2022). *Cannibal capitalism: How our system is devouring democracy, care, and the planet—and what we can do about it.* Verso.

Fraser, N., & Honneth, A. (2003). *Redistribution or recognition? A political–philosophical exchange.* Verso.

Fraser, N., & Jaeggi, R. (2018). *Capitalism: A conversation in critical theory.* Polity.

Fressoz, J.-B. (2015). Losing the Earth knowingly: Six environmental grammars around 1800. In C. Hamilton, C. Bonneuil, & F. Gemenne (Eds), *The Anthropocene and the global environmental crisis: Rethinking modernity in a new epoch* (pp. 70–84). Routledge.

Freud, S. (1953). The interpretation of dreams. In J. Strachey (Ed. & Trans.), *The standard edition of the complete psychological works of Sigmund Freud* (Vol. 4, pp. 1–630). Hogarth. (Original work published 1900)

Freud, S. (1955a). A difficulty in the path of psycho-analysis. In J. Strachey (Ed. & Trans.), *The standard edition of the complete psychological works of Sigmund Freud* (Vol. 17, pp. 135–144). Hogarth. (Original work published 1917)

Freud, S. (1955b). From the history of an infantile neurosis. In J. Strachey (Ed. & Trans.), *The standard edition of the complete psychological works of Sigmund Freud* (Vol. 17, pp. 7–124). Hogarth. (Original work published 1918)

Freud, S. (1955c). General theory of neuroses. In J. Strachey (Ed. & Trans.), *The standard edition of the complete psychological works of Sigmund Freud* (Vol. 17, pp. 243–463). Hogarth. (Original work published 1917)

Freud, S. (1955d). Studies on hysteria. In J. Strachey (Ed. & Trans.), *The standard edition of the complete psychological works of Sigmund Freud* (Vol. 2, pp. 19–182). Hogarth. (Original work published 1895)

Freud, S. (1955e). Totem and taboo. In J. Strachey (Ed. & Trans.), *The standard edition of the complete psychological works of Sigmund Freud* (Vol. 13, pp. 1–162). Hogarth. (Original work published 1913)

Freud, S. (1957a). Leonardo da Vinci and a memory of his childhood. In J. Strachey (Ed. & Trans.), *The standard edition of the complete psychological works of Sigmund Freud* (Vol. 11, pp. 63–138). Hogarth. (Original work published 1910)

Freud, S. (1957b). Thoughts for times on war and death. In J. Strachey (Ed. & Trans.), *The standard edition of the complete psychological works of Sigmund Freud* (Vol. 14, pp. 275–300). Hogarth. (Original work published 1915)

Freud, S. (1957c). The unconscious. In J. Strachey (Ed. & Trans.), *The standard edition of the complete psychological works of Sigmund Freud* (Vol. 14, pp. 159–215). Hogarth. (Original work published 1916)

Freud, S. (1958a). Notes on the unconscious in psychoanalysis. In J. Strachey (Ed. & Trans.), *The standard edition of the complete psychological works of Sigmund Freud* (Vol. 12, pp. 255–266). Hogarth. (Original work published 1912)

Freud, S. (1958b). On beginning treatment. In J. Strachey (Ed. & Trans.), *The standard edition of the complete psychological works of Sigmund Freud* (Vol. 12, pp. 121–144). Hogarth. (Original work published 1913)

Freud, S. (1958c). Remembering, repeating, and working through. In J. Strachey (Ed. & Trans.), *The standard edition of the complete psychological works of Sigmund Freud* (Vol. 12, pp. 145–156). Hogarth. (Original work published 1914)

Freud, S. (1959a). An autobiographical study. In J. Strachey (Ed. & Trans.), *The standard edition of the complete psychological works of Sigmund Freud* (Vol. 20, pp. 7–74). Hogarth. (Original work published 1925)

Freud, S. (1959b). Inhibitions, symptoms, and anxiety. In J. Strachey (Ed. & Trans.), *The standard edition of the complete psychological works of Sigmund Freud* (Vol. 20, pp. 87–178). Hogarth. (Original work published 1926)

Freud, S. (1959c). Question of lay analysis. In J. Strachey (Ed. & Trans.), *The standard edition of the complete psychological works of Sigmund Freud* (Vol. 20, pp. 183–258). Hogarth. (Original work published 1926)

Freud, S. (1961a). The future of an illusion. In J. Strachey (Ed. & Trans.), *The standard edition of the complete psychological works of Sigmund Freud* (Vol. 21, pp. 5–58). Hogarth. (Original work published 1927)

Freud, S. (1961b). Civilization and its discontents. In J. Strachey (Ed. & Trans.), *The standard edition of the complete psychological works of Sigmund Freud* (Vol. 21, pp. 64–148). Hogarth. (Original work published 1930)

Freud, S. (1962). Aetiology of hysteria. In J. Strachey (Ed. & Trans.), *The standard edition of the complete psychological works of Sigmund Freud* (Vol. 3, pp. 191–224). Hogarth. (Original work published 1897)

Freud, S. (1963). *Dora: An analysis of a case of hysteria*. Collier.

Freud, S. (1964a). Moses and monotheism. In J. Strachey (Ed. & Trans.), *The standard edition of the complete psychological works of Sigmund Freud* (Vol. 23, pp. 7–140). Hogarth. (Original work published 1939)

Freud, S. (1964b). My contact with Josef Popper-Lynkeus. In J. Strachey (Ed. & Trans.), *The standard edition of the complete psychological works of Sigmund Freud* (Vol. 22, pp. 219–224). Hogarth. (Original work published 1932)

Freud, S. (1964c). New introductory lectures on psycho-analysis. In J. Strachey (Ed. & Trans.), *The standard edition of the complete psychological works of Sigmund Freud* (Vol. 22, pp. 5–183). Hogarth. (Original work published 1933)

Freud, S. (1964d). Why war? In J. Strachey (Ed. & Trans.), *The standard edition of the complete psychological works of Sigmund Freud* (Vol. 22, pp. 199–205). Hogarth. (Original work published 1933)

Freud, S. (1999). Indexes and bibliographies. In A. Richards (Comp.), *The standard edition of the complete psychological works of Sigmund Freud.* Hogarth.

Freyd, J. J. (1996). *Betrayal trauma: The logic of forgetting childhood abuse.* Harvard University Press.

Fromm, E. (1973). *The anatomy of human destructiveness.* Open Road.

Fromm, E. (1977). *To have or to be?*Continuum.

Funk, R. (2024). On the psychodynamics of right-wing populism: A Frommian perspective. *Psychoanalytic Inquiry*, 44, 37–44. doi:10.1080/07351690.2023.2296341.

Gabbard, G. O. (1994). *Psychodynamic psychiatry in clinical practice.* American Psychiatric Press.

Gauthier, D. J. (2011). *Martin Heidegger, Emmanuel Levinas, and the politics of dwelling*. Lexington Books.

Gay, P. (1988). *Freud: A life for our time.* W. W. Norton.

Gentile, K. (2018). Animals as the symptom of psychoanalysis or, the potential for interspecies co-emergence in psychoanalysis. *Studies in Gender and Sexuality*, 19, 7–13.

Gentile, K. (2021). Kittens in the clinical space: Expanding subjectivity through dense temporalities of interspecies transcorporeal becoming. *Psychoanalytic Dialogues*, 31, 135–150.

Gentile, K. (2023). The irresistible lure of the fetus, or why abortion has everything to do with the colonizing temporalities of anti-blackness, "human" exceptionalism, and the climate crisis. *Studies in Gender and Sexuality*, 24, 43–52.

George, A. (Trans.). (2003). *The epic of Gilgamesh.* Penguin.

Gergely, G., & Unoka, Z. (2008). Attachment and mentalization in humans: The development of the affective self. In E. L. Jurist, A. Slade, & S. Bergner (Eds.) *Mind to mind: Infant research, neuroscience, and psychoanalysis* (pp. 50–87). Other Press.

Gherovici, P., & Christian, C. (2019). *Psychoanalysis in the barrios: Race, class, and the unconscious.* Routledge.

Gibson, N. C., & Beneduce, R. (2017). *Frantz Fanon, psychiatry, and politics.* Rowman & Littlefield.

Gilens, M., & Page, B. (2014). Testing theories of American politics: Elites, interest groups and average citizens. *American Political Science Association*, 12(3), 564–581.

Gill, M. M. (1982). *Analysis of transference: Vol. I—theory and technique.* International Universities Press.

Gilligan, C. (1982). *In a different voice: Psychological theory and women's development.* Harvard University Press.

Gilman, S. L. (1985a). *Difference and pathology: Stereotypes of sexuality, race, and madness.* Cornell University Press.

Gilman, S. L. (1985b). *Picturing health and illness: Images of identity and difference.* Johns Hopkins University Press.

Gilman, S. L. (1991). *Inscribing the other.* University of Nebraska Press.
Giovacchini, P. L. (1993). *Countertransference triumphs and catastrophes.* Aronson.
Giroux, H. A. (2012a). *Disposable youth: Racialized memories and the culture of cruelty.* Routledge.
Giroux, H. A. (2012b). *Zombie politics and culture in the age of casino capitalism.* Peter Lang.
Go, J. (2016). *Postcolonial thought and social theory.* Oxford University Press.
Goggin, J. E., & Goggin, E. B. (2001). *Death of a "Jewish science": Psychoanalysis in Nazi Germany.* Purdue University Press.
Goodall, J. (2010). *In the shadow of man.* Mariner.
Gordon, L. R. (2015). *What Fanon said: A philosophical introduction to his life and thought.* Fordham University Press.
Gordon, L. R. (2023). *Black existentialism and decolonizing knowledge.* Bloomsbury.
Grand, S. (1997). On the gendering of traumatic dissociation: A case of mother–son incest. *Gender and Psychoanalysis*, 2, 55–77.
Grand, S. (2000). *The reproduction of evil.* Analytic Press.
Gray, J. (2013). *The silence of animals: On progress and other modern myths.* Farrar, Straus, & Giroux.
Gray, J. G. (1970). *The warriors: Reflections on men in battle.* University of Nebraska.
Grayling, A. C. (2019). *The history of philosophy.* Penguin.
Green, A. (1999). *The dead mother: The work of André Green* (G. Kohon, Ed.). Routledge.
Greenberg, J. R., & Mitchell, S. A. (1983). *Object relations in psychoanalytic theory.* Harvard University Press.
Greenway, J. (2024). *Capitalism: A horror story—Gothic Marxism and the dark side of the radical imagination.* Repeater.
Gregory the Great. (1978). *Pastoral care.* Newman Press.
Groys, B. (2022). *Philosophy of care.* Verso.
Gutiérrez, G. (1985). *A theology of liberation: History, politics, and salvation.* Orbis Books.
Hale, N. G., Jr. (Ed.). (1971). *James Jackson Putnam and psychoanalysis: Letters between Putnam and Sigmund Freud, Ernest Jones, William James, Sandor Ferenczi, and Morton Prince, 1877–1917* (J. B. Heller, Trans.). Harvard University Press.
Hall, S. (Ed.). (1997). *Representation: Cultural representations and signifying practices.* Sage.
Hall, S. (2016). *Cultural studies 1983* (J. D. Slack & L. Grossberg, Eds.). Duke University Press.
Halpern, P. (2021). *Flashes of creation: George Gamow, Fred Hoyle, and the great Big Bang debate.* Basic Books.
Hamington, M. (2004). *Embodied care: Jane Addams, Maurice Merleau-Ponty, and feminist ethics.* University of Illinois.
Hamington, M. (2024). *Revolutionary care.* Routledge.
Han, B.-C. (2018a). *Saving beauty* (D. Steuer, Trans.). Polity.
Han, B.-C. (2018b). *Topology of violence* (A. Demarco, Trans.). MIT Press.
Han, B.-C. (2019). *What is power?*Polity.
Han, B.-C. (2022). *Non-things: Upheaval in the Lifeworld* (D. Steuer, Trans.). Polity.
Hanly, C. (2020). Psychoanalytic epistemology: Kant and Freud. *Psychoanalytic Quarterly*, 89, 305–337.

Harman, C. (2017). *A people's history of the world: From the Stone Age to the new millennium.* Verso.

Harrison, F. V. (Ed.). (1997). *Decolonizing anthropology: Moving further toward an anthropology for liberation.* American Anthropological Association.

Hart, D. B. (2024). *All things are full of gods: The mysteries of mind and life.* Yale University Press.

Harvey, D. (2016). *The ways of the world.* Oxford University Press.

Held, V. (Ed.). (1995). *Justice and care: Essential readings in feminist ethics.* Westview Press.

Held, V. (2006). *The ethics of care: Personal, political, and global.* Oxford University Press.

Herman, J. (1992). *Trauma and recovery: The aftermath of violence—from domestic abuse to political terror.* Basic Books.

Hillman, J. (2003). *A terrible love of war.* Penguin.

Hirsch, I. (1983). Analytic intimacy and the restoration of nurturance. *American Journal of Psychoanalysis*, 43, 325–343.

Hoggett, P. (2010). Government and the perverse social defence. *British Journal of Psychotherapy*, 26, 202–212.

Hoggett, P. (2012). Climate change in a perverse culture. In S. Weintrobe (Ed.), *Engaging with climate change: Psychoanalytic and interdisciplinary perspectives* (pp. 56–71). Routledge.

Hoggett, P. (2013). Governance and social anxieties. *Organizational and Social Dynamics*, 13, 69–78.

Hoggett, P. (2019). *Climate psychology: On indifference to disaster.* Palgrave Macmillan.

Hoggett, P. (2023). Reactionary states of mind as the Holocene ends. *Psychoanalytic Inquiry*, 43, 119–129. doi:10.1080/07351690.2023.2163149.

Holifield, E. B. (1983). *A history of pastoral care in America.* Wipf & Stock.

Holmes, J. (1996). *Attachment, intimacy, autonomy: Using attachment theory in adult psychotherapy.* Aronson.

Honneth, A. (1995). *The struggle for recognition.* MIT Press.

Honneth, A. (2023). Second nature: The profound depth of a key philosophical concept. In M. Hartman & A. Särkelä (Eds), *Naturalism and social philosophy* (pp. 19–34). Rowman & Littlefield.

Horner, A. J. (2005). *Dealing with resistance in psychotherapy.* Aronson.

Ibsen, H. (2009). *An enemy of the people, The wild duck, Rosmersholm.* Oxford University Press.

Ibsen, M. F. (2023). *A critical theory of global justice: The Frankfurt School and world society.* Oxford University Press.

Ingold, T. (2013). *Making: Anthropology, archeology, art, and architecture.* Routledge.

Ingold, T. (2022). *The perception of the environment: Essays on livelihood, dwelling and skill.* Routledge.

James, W. (1956). *The principles of psychology, vol.1. Henry Holt.* (Original work published 1918)

Janoff-Bulman, R. (1992). *Shattered assumptions: Towards a new psychology of trauma.* Free Press.

Johnson, M. (1987). *The body in the mind: The bodily basis of meaning, imagination, and reason.* University of Chicago Press.

Kahr, B. (1996). *D. W. Winnicott: A biographical portrait.* International Universities Press.

Kant, I. (1980). *Lectures on ethics.* Hackett.

Kassouf, S. (2017). Psychoanalysis and climate change: Revisiting Searles's *The Non-human Environment*, rediscovering Freud's phylogenetic fantasy, and imagining a future. *American Imago*, 74(2), 141–171. doi:10.1353/aim.2017.0008.

Kassouf, S. (2022). Thinking catastrophic thoughts: A traumatized sensibility on a hotter planet. *American Journal of Psychoanalysis*, 82, 60–79. doi:10.1057/s11231-11022-09340-09343.

Kaufmann, W. (1978). *Nietzsche: Philosopher, psychologist, antichrist.* Princeton University Press.

Kaur, V. (2020). *See no stranger: A memoir and manifesto of revolutionary love.* One World.

Kegley, J. A. K. (1980). Josiah Royce on self and community. *Rice University Studies*, 66(4), 33–53. doi:10.4000/ejpap.3123.

Keown, D. (2000). *Contemporary Buddhist ethics.* Routledge.

Kernberg, O. (1984). *Object relations theory and clinical psychoanalysis.* Aronson.

Kerven, R. (2018). *Native American myths: Collected 1636–1919.* Talking Stone.

Kestenberg, J., & Weinstein, J. (1978). Transitional objects and body image formation. In S. A. Grolnick & L. Barkin (Eds.), *Between reality and fantasy: Transitional objects and phenomena* (pp. 75–96). Aronson.

Khan, M. M. R. (1963). The concept of cumulative trauma. *Psychoanalytic Study of the Child*, 18, 286–306. doi:10.1080/00797308.1963.11822932.

Khurana, T. (2023). The stage of difference: On the second nature of civil society in Kant and Hegel. In M. Hartman & A. Särkelä (Eds.), *Naturalism and social philosophy: Contemporary perspectives* (pp. 35–64). Rowman & Littlefield.

Klein, N. (2014). *This changes everything: Capitalism vs. the climate.* Simon & Schuster.

Kniess, J. (2018). Bentham on animal welfare. *British Journal for the History of Philosophy*, 27(3), 556–572. doi:10.1080/09608788.2018.1524746.

Kohn, E. (2013). *How forests think: Toward an anthropology beyond the human.* University of California Press.

Kohut, H. (1984). *How does analysis cure?*Chicago University Press.

Kolbert, E. (2014). *The sixth extinction: An unnatural history.* Henry Holt.

Kompridis, N. (2020). Nonhuman agency and human normativity. In A. Bilgrami (Ed.), *Nature and value* (pp. 240–260). Columbia University Press.

Kotsko, A. (2020). *Agamben's philosophical trajectory.* University of Edinburgh Press.

Kovel, J. (1970). *White racism: A psychohistory.* Columbia University Press.

Kovel, J. (1984). On being a Marxist psychoanalyst (and a psychoanalytic Marxist). *Free Associations*, 1, 149–154.

Kovel, J. (1988). *The radical spirit: Essays on psychoanalysis and society.* Free Associations Books.

Kumin, I. (1996). *Pre-object relatedness: Early attachment and the psychoanalytic situation.* Guilford.

Laing, R. D. (1969). *The divided self: An existential study in sanity and madness.* Penguin.

Lakoff, G., & Johnson, M. (1999). *Philosophy in the flesh: The embodied mind and its challenge to Western thought.* Basic Books.

LaMothe, R. (1999). The absence of cure: The core of malignant trauma and symbolization. *Journal of Interpersonal Violence*, 14, 1193–1210. doi:10.1177/088626099014011005.

LaMothe, R. (2004). Freud's envy of religious experience. *International Journal for the Psychology of Religion*, 14(3), 161–176. doi:10.1207/s15327582ijpr1403_2.

LaMothe, R. (2014). Winnicott and helplessness: Developmental theory, religion, and personal life. *Psychoanalytic Quarterly*, 83(4), 871–896. doi:10.1002/j.2167-4086.2014.00125.x.

LaMothe, R. (2017). *Care of souls, care of polis: Toward a political pastoral theology.* Cascade.

LaMothe, R. (2021). Illusions, political selves, and responses to the Anthropocene Age: A political-psychoanalytic perspective. *Free Associations*, 22, 1–18. doi:10.1234/fa.v0i84.409.

LaMothe, R. (2023). Decolonizing nature: Freud, the ontological rift, and the Anthropocene Age. *Free Associations: Psychoanalysis and Culture, Media, Groups, Politics*, 90. https://freeassociations.org.uk/FA_New/OJS/index.php/fa/article/view/469.

LaMothe, R. (2024). *A political psychoanalysis for the Anthropocene Age: The fierce urgency of now.* Routledge.

LaMothe, R., Arnold, J., & Crane, J. (1998). The penumbra of religious discourse. *Psychoanalytic Psychology*, 15(1), 63–73. doi:10.1037/0736-9735.15.1.63.

Langs, R. (1989). *The technique of psychoanalytic therapy.* Aronson.

Latour, B. (1993). *We have never been modern.* Harvard University Press.

Latour, B. (1996). Not the question. *Anthropology Newsletter*, 37(3), 1–5. doi:10.1111/an.1996.37.3.1.2.

Laubscher, L., Hook, D., & Desai, M. U. (Eds.). (2022). *Fanon, phenomenology, and psychology.* Routledge.

Laymon, K. (2019). *Heavy: An American memoir.* Scribner.

Layton, L. (2020). *Toward a social psychoanalysis: Culture, character, and normative unconscious processes.* Routledge.

Layton, L., Hollander, N., & Gutwill, S. (Eds.). (2006). *Psychoanalysis, class, and politics: Encounters in the clinical setting.* Routledge.

Lear, J. (1998). *Open minded: Working out the logic of the soul.* Harvard University Press.

Lear, J. (2006). *Radical hope: Ethics in the face of cultural devastation.* Harvard University Press.

Lechte, J., & Newman, S. (2015). *Agamben and the politics of human rights: Statelessness, images, violence.* Edinburgh University Press.

Léger-Goodes, T., Malboeuf-Hurtubise, C., Mastine, T., Généreux, M., Paradis, P.-O., & Camden, C. (2022). Eco-anxiety in children: A scoping review of the mental health impacts of the awareness of climate change. *Frontiers in Psychology*, 13, Article 872544. doi:10.3389/fpsyg.2022.872544.

Lentz, J. (2016). Reconsidering resistance. *Psychoanalytic Psychology*, 33, 599–609. doi:10.1037/a0038918.

Leone, M. (2020). *On insignificance: The loss of meaning in the post-material age.* Routledge.

Lepore, J. (2018). *These truths: A history of the United States.* Norton.

Lertzman, R. (2017). *Environmental melancholia: Psychoanalytic dimensions of engagement.* Routledge.

Levi, P. (1960). *Survival in Auschwitz*. Collier.

Levin, F. M., & Trevarthen, C. (2000). Subtle is the Lord: The relationship between consciousness, the unconscious, and the executive control network (ECN) of the brain. *Annual of Psychoanalysis*, 28, 105–125.

Levinas, E. (1969). *Totality and infinity: An essay on exteriority.* Duquesne University Press.
Levine, L. (2009a). Impasse and resonance across multiple relational realms: Reply to commentaries. *Psychoanalytic Dialogues*, 19, 480–485.
Levine, L. (2009b). Transformative aspects of our own analyses and their resonance in our work with our patients. *Psychoanalytic Dialogues*, 19, 454–462. doi:10.1080/10481880903088641.
Linzey, A. (2009). *Creatures of the same God: Explorations in animal theology.* Lantern.
Linzey, A. (2013). *Why animal suffering matters: Philosophy, theology, and practical ethics.* Oxford University Press.
Litt, C. J. (1986). Theories of transitional object attachment: An overview. *International Journal of Behavioral Health*, 9, 383–399. doi:10.1177/016502548600900308.
Lloyd, D. (2024). *Land is kin: Sovereignty, religious freedom, and Indigenous sacred sites.* University of Kansas Press.
Loewald, H. W. (2000). *The essential Loewald: Collected papers and monographs.* University Publishing Group.
Løgstrup, K. E. (1997). *The ethical demand.* Notre Dame University Press.
London, J. (1981). *Call of the wild.* Running Press. (Original work published 1903)
Lopez, B. (2022). *Embrace fearlessly the burning world.* Random House.
Lukács, G. (1968). *History and class consciousness: Studies in Marxist dialectics.* MIT Press.
Lyotard, J.-F. (1999). *The postmodern condition: A report on knowledge.* University of Minneapolis Press.
MacKinnon, C. A. (1991). *Toward a feminist theory of the state.* Harvard University Press.
Macmurray, J. (1961). *Persons in relation.* Humanities Press International.
Macmurray, J. (1993). *Conditions of freedom.* Humanities Press International. (Original work published 1949)
Malpas, J. (2006). *Heidegger's topology: Being, place, world.* MIT Press.
Mander, J. (2012). *The capitalism papers: Fatal flaws of an obsolete system.* Counterpoint Press.
Marcuse, H. (1964). *One-dimensional man: Studies in the ideology of advanced industrial society.* Beacon Press.
Margalit, A. (1996). *The decent society.* Harvard University Press.
Margulis, L. (1997). *Microcosmos: Four billion years of microbial evolution.* University of California Press.
Margulis, L. (2007). *Dazzle gradually: Reflections on the nature of nature.* Chelsea Green.
McDonough, K. S. (2024). *Indigenous science and technology: Nahuas and the world around them.* University of Arizona Press.
McFarlane, A. C., & van der Kolk, B. A. (1996). Trauma and its challenge to society. In B. A. van der Kolk, A. C. McFarlane, & L. Weisaeth (Eds.), *Traumatic stress: The effects of overwhelming experience on mind, body, and society* (pp. 24–46). Guilford.
McGuire, D. L. (2011). *At the dark end of the street: Black women, rape, and resistance—a new history of the civil rights movement from Rosa Parks to the rise of black power.* Random House.
McGuire, W. (1974). *The Freud/Jung letters: The correspondence between Sigmund Freud and C. G. Jung.* Princeton University Press.

McLaren, M. A. (Ed.). (2017). *Decolonizing feminism: Transnational feminism and globalization*. Rowman & Littlefield.

Mei, T. S. (2017). *Land and the given economy: The hermeneutics and phenomenology of dwelling*. Northwestern University Press.

Meijer, E. (2019). *When animals speak: Toward an interspecies democracy.* New York University Press.

Meijer, E. (2020). *Animal languages* (L. Watkinson, Trans). MIT Press.

Meissner, W. W. (1992). *Ignatius of Loyola: The psychology of a saint*. Yale University Press.

Meister, M. (2024). Luxury after Doomsday: Billionaire bunkers, *Science Illustrated*, August 11, 2024. Available at: https://scienceillustrated.com/technology/construction/buildings/luxury-after-doomsday-billionaire-bunkers (accessed June 12, 2025).

Melville, H. (1851). *Moby-Dick*. E-Book.

Merleau-Ponty, M. (1964). *The primacy of perception and other essays on phenomological psychology, the philosophy of art, history and politics* (W. Cobb, Trans., J. M. Edie, Ed.). Northwestern University Press.

Messerly, J. G. (2014, May 22). W. H. Auden's "We must love one another or die." *Reason and Meaning*. Available at: https://reasonandmeaning.com/2014/05/22/w-h-audens-we-must-love-one-another-or-die/ (accessed January 19, 2026).

Mill, J. S. (1970). *On nature*. Taylor & Francis.

Miller, A. (2002). *For your own good: Hidden cruelty in child-rearing and the roots of violence*. Farrar, Straus, & Giroux.

Mills, C. W. (1997). *The racial contract*. Cornell University Press.

Mills, C. W. (2017). *Black rights/White wrongs.* Oxford University Press.

Moltmann, J. (1973). *The gospel of liberation*. Word Books.

Monsó, S. (2024). *Playing possum: How animals understand death*. Princeton University Press.

Montgomery, S. (2016). *The soul of an octopus: A surprising exploration into the wonder of consciousness.* Atria Books.

Moore, J. W. (2016, October 9). Name the system! Anthropocene and the Capitalocene alternative. Available at: https://jasonwmoore.wordpress.com/tag/capitalocene/.

Morris, C. (1964). *Signification and significance: A study of the relations of signs and values.* MIT Press.

Morton, T. (2017). *Humankind: Solidarity with nonhuman people.* Verso.

Morton, T. (2024). *Hell: In search of a Christian ecology.* Columbia University Press.

Murdoch, I. (2001). *The sovereignty of good.* Routledge.

Neihardt, J. G. (2014). *Black Elk speaks.* University of Nebraska Press. (Original work published 1932)

Newman, S. (2016). *Postanarchism.* Polity.

Newmeyer, S. T. (2005). *Animals, rights and reason in Plutarch and modern ethics.* Routledge.

Niebuhr, H. R. (1989). *Faith on earth: An inquiry into the structure of human faith.* Yale University Press.

Nietzsche, F. (1968). *The will to power* (W. Kaufmann & R. J. Hollingdale, Trans.). Vintage.

Nishitani, K. (1982). *Religion and nothingness* (J. Van Bragt, Trans.). University of California Press.

Noddings, N. (1984). *Caring: A feminine approach to ethics and moral education.* University of California Press.
Nussbaum, M. C. (2022). *Justice for animals: Our collective responsibility.* Simon & Schuster.
Oksala, J. (2012). *Foucault, politics, and violence.* Northwestern University Press.
Oliner, P. M., & Oliner, S. P. (1995). *Toward a caring society: Ideas into action.* Praeger.
Orange, D. M. (2009). *Thinking for clinicians: Philosophical resources for contemporary psychoanalysis and the humanistic psychotherapies.* Routledge.
Orange, D. M. (2017). *Climate crisis, psychoanalysis, and radical ethics.* Routledge.
Ortega y Gasset, J. (1957). *Man and people.* W. W. Norton.
Panetta, V. (2002). Fantasy as resistance. *Modern Psychoanalysis*, 27, 101–111.
Parker, I., & Pavón-Cuéllar, D. (2021). *Psychoanalysis and revolution: Critical psychology for liberation movements.* 1968 Press.
Patterson, O. (1982). *Slavery and social death: A comparative study.* Harvard University Press.
Peirce, C. S. (1991). *Peirce on signs: Writings on semiotic by Charles Sanders Peirce* (J. Hoopes, Ed.). University of North Carolina Press.
Penrose, R., Hameroff, S., & Kak, S. (Eds.). (2017). *Consciousness and the universe: Quantum physics, evolutions, brain & mind.* Cosmology Science.
Phillips, A. (1993). *On kissing, tickling, and being bored: Psychoanalytic essays on the unexamined life.* Harvard University Press.
Pihkala, P. (2017a). *Early ecotheology and Joseph Sittler.* LIT Verlag.
Pihkala, P. (2017b). Environmental education after sustainability: Hope in the midst of tragedy. *Global Discourse*, 7, 109–127. doi:10.1080/23269995.2017.1300412.
Pihkala, P. (2018a). Death, the environment, and theology. *Dialog*, 57, 287–294. doi:10.1111/dial.12437.
Pihkala, P. (2018b). Eco-anxiety, tragedy, and hope: Psychological and spiritual dimensions of climate change. *Zygon*, 53, 545–569. doi:10.1111/zygo.12407.
Pihkala, P. (2020a). Anxiety and the ecological crisis: An analysis of eco-anxiety and climate anxiety. *Sustainability*, 12, Article 7836. doi:10.3390/su12197836.
Pihkala, P. (2020b). Eco-anxiety and environmental education. *Sustainability*, 12, Article 10149.
Pihkala, P. (2022a). Eco-anxiety and pastoral care: Theoretical considerations and practical suggestions. *Religions*, 13, 1–19. doi:10.3390/su122310149.
Pihkala, P. (2022b). The process of eco-anxiety and ecological grief: A narrative review and a new proposal. *Sustainability*, 14(24), 1–53. doi:10.3390/su142416628.
Pihkala, P. (2022c). Toward a taxonomy of climate emotions. *Frontiers in Climate*, 3. doi:10.3389/fclim.2021.738154.
Pihkala, P. (2024). Ecological sorrow: Types of grief and loss in ecological grief. *Sustainability*, 16(2), Article 849. doi:10.3390/su16020849.
Piketty, T. (2014). *Capital in the twenty-first century.* Belknap.
Piketty, T. (2020). *Capital and ideology.* Harvard University Press.
Plato. (1956). *The great dialogues of Plato* (W. H. D. Rouse, Trans.). Mentor Books.
Polkinghorne, D. E. (1988). *Narrative knowing and the human sciences.* SUNY.
Popper, K. (2002). *The open society and its enemies.* Routledge.
Porphyry. (1823). *On abstinence from animal food.* Book 1 (pp. 11–44). Available at: www.tertullian.org/fathers/porphyry_abstinence_01_book1.htm.

Porter, R. (1991a). *The Faber book of madness.* Faber.

Porter, R. (1991b). *A social history of madness: The world through the eyes of the insane.* Dutton.

Prichard, A. (2022). *Anarchism: A very short introduction.* Oxford.

Prozorov, S. (2014). *Agamben and politics: A critical introduction.* Edinburgh University Press.

Pruyser, P. W. (1983). *The play of the imagination: Towards a psychoanalysis of culture.* International Universities Press.

Puryear, S. (2017). Schopenhauer on the rights of animals. *European Journal of Philosophy*, 25(2), 250–269. doi:10.1111/ejop.12237.

Rank, O. (2014). *The trauma of birth.* Routledge. (Original work published 1929)

Rendón, M. (2020). The morality of evolution and a return to subjectivity. *American Journal of Psychoanalysis*, 80, 1–15. doi:10.1057/s11231-11020-09231-09235.

Ricoeur, P. (1970). *Freud and philosophy: An essay on interpretation.* Yale University Press.

Rieff, P. (2006). *Triumph of the therapeutic: Uses of faith after Freud.* Intercollegiate Studies.

Rigby, K. (2017). Religion and ecology: Towards a communion of creatures. In S. Oppermann & S. Iovino (Eds.), *Environmental humanities: Voices from the Anthropocene* (pp. 273–294). Rowman & Littlefield.

Robinson, C. (1983). *Black Marxism: The making of the black radical tradition.* University of North Carolina Press.

Robinson, F. (1999). *Globalizing care: Ethics, feminist theory, and international relations.* Westview Press.

Robinson, F. (2011). *The ethics of care: A feminist approach to human security.* Temple University Press.

Rodman, F. R. (2003). *Winnicott: Life and work.* Da Capo Press.

Rosa, H. (2019). *Resonance: A sociology of our relationship to the world* (J. C. Wagner, Trans.). Polity.

Rosa, H. (2020). *The uncontrollability of the world.* Polity.

Rossing, B. R. (2022). Waters cry out: Water protectors, watershed justice, and the voice of waters. In S. L. Mendoza & G. Zachariah (Eds.), *Decolonizing ecotheology* (pp. 39–57). Pickwick.

Rousseau, B. (2016, July 13). In New Zealand, lands and rivers can be people (legally speaking). *New York Times.* Available at: www.nytimes.com/2016/07/14/world/what-in-the-world/in-new-zealand-lands-and-rivers-can-be-people-legally-speaking.html.

Rousseau, J.-J. (2016). *Discourse on inequality.* Sovereign.

Royce, J. (2018). *The philosophy of loyalty.* CreateSpace Independent Publishing Platform. (Original work published 1908)

Runciman, D. (2023, November 2). The Leviacene: Defining our times (No. 28) [Audio podcast episode]. In *Past Present Future.* Available at: https://podcasts.apple.com/us/podcast/the-leviacene-defining-our-times/id1682047968?i=1000633437790.

Ryan, A. (2012). *On politics: A history of political thought: From Herodotus to the present.* Liveright.

Saïd, E. W. (1979). *Orientalism.* Vintage Books.

Saïd, E. W. (1994). *Culture and imperialism.* Vintage Books.

Samberg, E. (2004). Resistance: How do we think of it in the twenty-first century? *Journal of the American Psychoanalytic Association*, 52, 243–253.

Samuels, A. (1993). *The political psyche*. Routledge.
Samuels, A. (2001). *Politics on the couch: Citizenship and the internal life*. Karnac Books.
Samuels, A. (2004). Politics on the couch? Psychotherapy and society—some possibilities and limitations. *Psychoanalytic Dialogues*, 14, 817–834. doi:10.1080/10481881409348809.
Samuels, A. (2015). *A new therapy for politics?*Karnac Books.
Samuels, R. (2023). *Psychoanalysis and the future of global politics: Overcoming climate change, pandemics, war, and poverty*. Palgrave Macmillan.
Sartre, J.-P. (1965). *Essays in existentialism*. Citadel.
Sass, L. (1992). *Madness and modernism: Insanity in the light of modern art, literature, and thought*. Basic Books.
Schlanger, Z. (2024). *The light eaters: How the unseen world of plant intelligence offers a new understanding of life on Earth*. Harper.
Schmitt, C. (2005). *Political theology: Four chapters on the concept of sovereignty* (G. Schwab, Trans.). University of Chicago Press.
Schoenewolf, G. (1993). *Counterresistance: The therapist's interference with the therapeutic process*. Aronson.
Scull, A. (2015). *Madness in civilization: A cultural history of insanity from the Bible to Freud, from madhouse to modern medicine*. Princeton University Press.
Scully, M. (2003). *Dominion: The power of man, the suffering of animals, and the call to mercy*. St. Martin's Griffin.
Searles, H. F. (1960). *The nonhuman environment in normal development and schizophrenia*. International Universities Press.
Segal, H. (1957). Notes on symbol formation. *International Journal of Psychoanalysis*, 38, 391–397.
Segal, L. (2023). *Lean on me: A politics of radical care*. Verso.
Sehgal, P. (2022, January 3). The case against the trauma plot. *The New Yorker*, 1–9. Available at: www.newyorker.com/magazine/2022/01/03/the-case-against-the-trauma-plot.
Self, W. (2022). How everything became trauma. *Harper's Magazine*, 1–15. Available at: https://harpers.org/archive/2021/12/a-posthumous-shock-trauma-studies-modernity-how-everything-became-trauma/.
Sevenhuijsen, S. (1998). *Citizenship and the ethics of care: Feminist considerations on justice, morality and politics*. Routledge.
Shaw, D. (2003). On the therapeutic action of analytic love. *Contemporary Psychoanalysis*, 39, 251–278. doi:10.1080/00107530.2003.10745820.
Sheehan, S. (2003). *Anarchism*. Reaktion Books.
Sheehi, L., & Sheehi, S. (2022). *Psychoanalysis under occupation: Practicing resistance in Palestine*. Routledge.
Sheehi, S. (2018). The transnational Palestinian self: Toward decolonizing psychoanalytic thought. *Psychoanalytic Perspectives*, 15, 307–322. doi:10.1080/1551806X.2018.1492816.
Shelley, M. (2000). *Frankenstein*. Signet Classic.
Sherman, M. H. (2007). Siding with the resistance in paradigmatic psychotherapy. *Modern Psychoanalysis*, 32, 177–193.
Simek, N. (2011). Postcolonial Freud: Psychoanalysis in the French Antilles. *Psychoanalysis and History, 13*, 227–243. doi:10.3366/pah.2011.0090.
Singer, P. (1975). *Animal liberation*. Harper Collins.

Singer, P. (2023). *Ethics in the real world: 90 essays on things that matter.* Princeton University Press.

Slipp, S. (1991). *The technique and practice of object relations family theory.* Aronson.

Smith, D. L. (2021). *Making monsters: The uncanny power of dehumanization.* Harvard University Press.

Sokolowsky, L. (2022). *Psychoanalysis under Nazi occupation: The origins, impact and influence of the Berlin Institute.* Routledge.

Sollereder, B. N. (2020). *God. evolution, and animal suffering: Theodicy without a fall.* Routledge.

Solomon, A. (2015). *The noonday demon: An atlas of depression.* Scribner.

Soss, J., Fording, R. C., & Sanford, S. F. (2011). *Disciplining the poor: Neoliberal paternalism and the persistent power of race.* University of Chicago Press.

Spivak, G. C. (2004). *Can the subaltern speak?*Routledge.

Stengers, I. (2023). *Making sense in common: A reading of Whitehead in times of collapse.* University of Minnesota Press.

Stern, D. N. (1997). *Unformulated experience: From dissociation to imagination in psychoanalysis.* Analytic Press.

Stone, M. (2006). The analyst's body as tuning fork: Embodied resonance in countertransference. *Journal of Analytical Psychology,* 51, 109–124. doi:10.1111/j.1465-5922.2006.575_1.x.

Stonebridge, L. (2024). *We are free to change the world: Hannah Arendt's lessons in love and disobedience.* Hogarth.

Strenger, C. (2011). *The fear of insignificance: Searching for meaning in the twenty-first century.* Palgrave Macmillan.

Sulloway, F. J. (1992). *Freud, Biologist of the mind: Beyond the psychoanalytic legend.* Harvard University Press.

Sung, J.-M. (2007). *Desire, market and religion.* SCM Press.

Swartz, S. (2018). Counter-recognition in decolonial struggle. *Psychoanalytic Dialogues,* 28, 520–527. doi:10.1080/10481885.2018.1506227.

Szasz, T. S. (1977). *The manufacture of madness: A comparative study of the Inquisition and the mental health movement.* Harper Torchbooks.

Szasz, T. S. (1987). *Insanity: The idea and its consequences.* John Wiley.

Tansey, M. J., & Burke, W. F. (1989). *Understanding countertransference: From projective identification to empathy.* Analytic Press.

Taylor, C. (2024). *Cosmic connections: Poetry in the age of disenchantment.* Belknap.

Taylor, D. (2023, July 31). How humans will become a multi-planetary species. *Newsweek.* Available at: www.newsweek.com/how-humans-will-become-multi-planetary-species-1816016 (accessed December 4, 2025).

Teilhard de Chardin, P. (1978). *Activation of energy: Enlightening reflections on spiritual energy* (R. Hague, Trans.). Harcourt.

Thomä, H., & Kächele, H. (1994). *Psychoanalytic practice.* Aronson.

Thorpe, C. R. (2020). Escape from reflexivity: Fromm and Giddens on individualism, anxiety, and authoritarianism. In K. Durkin & J. Braune (Eds.), *Erich Fromm's critical theory: Hope, humanism, and the future* (pp. 166–193). Bloomsbury.

Tippett, K. (2016, September 15). Ruby Sales—where does it hurt? (radio broadcast). *On Being with Krista Tippett.* Available at: https://onbeing.org/programs/ruby-sales-where-does-it-hurt/#transcript (last updated January 16, 2020, accessed December 4, 2025).

Tocqueville, A.de. (2004). *Democracy in America.* Bantam Classic.

Tonner, P. (2018). *Dwelling: Heidegger, archaeology, morality.* Routledge.

Trevarthen, C. (1993). Playing into reality: Conversations with the infant communicator. *Winnicott Studies*, 7, 67–84.

Tronto, J. (1993). *Moral boundaries: A political argument for an ethic of care.* Routledge.

Tronto, J. C. (2013). *Caring democracy: Markets, equality, and justice.* New York University Press.

Trosper, R. L. (2022). *Indigenous economics: Sustaining peoples and their lands.* University of Arizona Press.

Turnbull, O. H., & Bär, A. (2020). Animal minds: The case for emotion, based on neuroscience. *Neuropsychoanalysis*, 22, 109–128. doi:10.1080/15294145.2020.1848611.

Twine, R. (2024). *The climate crisis and other animals.* Sydney University Press.

Ugilt, R. (2014). *Giorgio Agamben: Political philosophy.* Humanities-Ebooks.

Valencia, S. (2018). *Gore capitalism* (J. Pluecker, Trans.). Semiotext(e).

Välimäki, S. (2015). Psychoanalysis, resonance, and the art of listening: A comment on Arnfinn Bø-Rygg's "Hearing, Listening, and the Voice." *Scandinavian Psychoanalytic Review*, 38, 152–155. doi:10.1080/01062301.2015.1104830.

Van der Kolk, B. (2015). *The body keeps the score: Brain, mind, and body in the healing of trauma.* Penguin.

Van der Kolk, B. A., van der Hart, O., & Marmar, C. R. (1996). Dissociation and information processing in posttraumatic stress disorder. In B. A. van der Kolk, A. C. McFarlane, & L. Weisaeth (Eds.), *Traumatic stress: The effects of overwhelming experience on mind, body, and society* (pp. 303–330). Guilford.

Van der Kolk, B. A., Weisaeth, L., & van der Hart, O. (1996). History of trauma in psychiatry. In B. van der Kolk, A. McFarlane, & L. Weisaeth (Eds.), *Traumatic stress: The effects of overwhelming experience on mind, body, and society* (pp. 47–76). Guilford.

Viveiros de Castro, E. (2017). *Cannibal metaphysics.* Minnesota University Press.

Volf, M. (1996). *Exclusion and embrace: A theological exploration of identity, otherness, and reconciliation.* Abingdon Press.

Wacquant, L. (2009). *Punishing the poor: The neoliberal government of social insecurity.* Duke University Press.

Waggoner, M. (2018). *Unhoused: Adorno and the problem of dwelling.* Columbia Books.

Watterson, B., & Trudeau, G. B. (1985–1995). *Calvin and Hobbes* [comic strip]. Andrews McMeel.

Weber, M. (1992). *The Protestant ethic and the spirit of capitalism.* Routledge.

Weintrobe, S. (2010). On links between runaway consumer greed and climate denial: A psychoanalytic perspective. *Bulletin of the British Psychoanalytic Society*, 1, 63–75. Institute of Psychoanalysis.

Weintrobe, S. (Ed.). (2013). *Engaging with climate change: Psychoanalytic and interdisciplinary perspectives.* Routledge.

Weintrobe, S. (2021). *The psychological roots of the climate crisis: Neoliberal exceptionalism and the culture of uncare.* Bloomsbury.

Weintrobe, S. (2022). The new bold imagination needed to repair and expand the ecological self. In W. Holloway, P. Hoggett, C. Robertson, & S. Weintrobe (Eds.), *Climate psychology: A matter of life and death* (pp. 97–120). Phoenix.

Weintrobe, S. (2024). Silencing is the real crime: Youth and elders talk about climate. *Psychoanalytic Study of the Child*, 77, 341–355.
Wilkerson, I. (2020). *Caste: The origins of our discontents.* Random House. doi:10.1080/00797308.2023.2287380.
Williams, L. M., & Banyard, V. L. (1999). *Trauma and memory.* Sage.
Wilson, E. O. (2003). *The future of life.* Vintage.
Winnicott, D. W. (1945). Primitive emotional development. *International Journal of Psychoanalysis*, 26, 137–143.
Winnicott, D. W. (1965). *The maturational processes and the facilitating environment.* The International Psychoanalytic Library (Vol. 64). Hogarth Press.
Winnicott, D. W. (1971). *Playing and reality.* Routledge.
Winnicott, D. W. (1975). *Through paediatrics to psycho-analysis.* The International Psychoanalytic Library (Vol. 100). Hogarth.
Wolfberg, L. R. (1995). *The technique of psychotherapy.* Aronson.
Wolin, S. S. (2008). *Democracy incorporated: Managed democracy and the specter of inverted totalitarianism.* Princeton University Press.
Wolin, S. S. (2016). *Fugitive democracy and other essays.* Princeton University Press.
Wolstein, B. (1995a). *Countertransference.* Aronson.
Wolstein, B. (1995b). *Transference: Its structure and function in psychoanalytic therapy.* Aronson.
Wolynn, M. (2017). *It didn't start with you: How inherited family trauma shapes who we are and how to end the cycle.* Penguin.
Woods, E. M. (2017). *The origin of capitalism: A longer view.* Verso.
Wu, J., Gaelen, S., & Hasina, S. (2020). Climate anxiety in young people: A call to action. *The Lancet Planetary Health*, 4(10), 435–436. doi:10.1016/S2542-5196(20)30223-0.
Wulf, A. (2015). *The invention of nature: Alexander von Humboldt's new world.* Vintage.
Yong, E. (2023). *An immense world: How animal senses reveal the hidden realms around us.* Random House.
Zachariah, G. (2024). Moana eco-theology: Toward an eco-theology of commoning. In S. Oppermann & S. Iovino (Eds.), *Environmental humanities: Voices from the Anthropocene* (pp. 65–82). Rowman & Littlefield.
Zaretsky, E. (1976). *Capitalism, the family, and personal life.* Harper.
Zaretsky, E. (2005). *Secrets of the soul: A social and cultural history of psychoanalysis.* Knopf Doubleday.
Zeddies, T. J. (2002). Behind, beneath, above, and beyond: The historical unconscious. *Journal of the American Academy of Psychoanalysis*, 30, 211–222. doi:10.1521/jaap.30.2.211.21954.
Zinn, H. (2003). *A people's history of the United States.* Harper Perennial.
Zizioulas, J. D. (1985). *Being as communion: Studies in personhood and the church.* St. Vladimir's Seminary Press.
Zizioulas, J. D. (2006). *Communion and otherness: Further studies in personhood and the church.* T&T Clark.
Zúñiga, D. (2023). *Pluralist politics, relational worlds: Vulnerability and care of the Earth.* University of Toronto Press.

Index

For Product Safety Concerns and Information please contact our EU representative GPSR@taylorandfrancis.com
Taylor & Francis Verlag GmbH, Kaufingerstraße 24, 80331 München, Germany

www.ingramcontent.com/pod-product-compliance
Lightning Source LLC
LaVergne TN
LVHW010857110826
845149LV00005B/1416

* 9 7 8 1 0 3 2 8 5 4 1 0 6 *